HAM RADIO

How to Setup and Operate Your Ham Radio

(Quickstart Guide for New Hams and Amateur Radio Enthusiasts)

Carlos Martinez

Published By Carlos Martinez

Carlos Martinez

All Rights Reserved

Ham Radio: How to Setup and Operate Your Ham Radio (Quickstart Guide for New Hams and Amateur Radio Enthusiasts)

ISBN 978-1-77485-426-6

All rights reserved. No part of this guide may be reproduced in any form without permission in writing from the publisher except in the case of brief quotations embodied in critical articles or reviews.

Legal & Disclaimer

The information contained in this book is not designed to replace or take the place of any form of medicine or professional medical advice. The information in this book has been provided for educational and entertainment purposes only.

The information contained in this book has been compiled from sources deemed reliable, and it is accurate to the best of the Author's knowledge; however, the Author cannot guarantee its accuracy and validity and cannot be held liable for any errors or omissions. Changes are periodically made to this book. You must consult your doctor or get professional medical advice before using any of the suggested remedies, techniques, or information in this book.

Upon using the information contained in this book, you agree to hold harmless the Author from and against any damages, costs, and expenses, including any legal fees potentially resulting from the application of any of the information provided by this guide. This disclaimer applies to any damages or injury caused by the use and application, whether directly or indirectly, of any advice or information presented, whether for breach of contract, tort, negligence, personal injury, criminal intent, or under any other cause of action.

You agree to accept all risks of using the information presented inside this book. You need to consult a professional medical practitioner in order to ensure you are both able and healthy enough to participate in this program.

TABLE OF CONTENTS

INTRODUCTION 1

CHAPTER 1: WHAT'S HAM RADIO STATION? 3

CHAPTER 2: WHAT TO PICK ANTENNAS FOR HAM RADIO 40

CHAPTER 3: WHAT TO PICK THE RIGHT COMPUTER TO RUN THE HAM RADIO STATION OF YOUR CHOICE. HAM RADIO STATION 54

CHAPTER 4: HOW TO ORGANIZE THE OPERATION OF YOUR HAM RADIO STATION 64

CHAPTER 5: WHAT TO DO KEEP THE LOG OF YOUR CONTACTS 76

CHAPTER 6: HOW TO OPERATE AWAY FROM HOME 84

CHAPTER 7: UNDERSTANDING HANDS-ON RADIO 89

CHAPTER 8: THE BASIC TECHNICAL INFORMATION REQUIRED FOR HAM RADIO STATION 98

CHAPTER 9: HOW TO MAKE CONTACT 106

CHAPTER 10: UNDERSTANDING QSOS (CONTACTS) 115

CHAPTER 11: INVOLVING IN CASUAL OPERATION 121

CHAPTER 12: PARTICIPATING PUBLIC SERVICE OPERATIONS PUBLIC SERVICE OPERATIONS 129

CHAPTER 13: UNDERSTANDING DIGITAL DEFINITIONS .. 137

CHAPTER 14: HOW TO PARTICIPATE IN RADIO CONTESTS 143

CHAPTER 15: UNDERSTANDING THE MORSE CODE AND USING MORSE CODE 147

CHAPTER 16: UNDERSTANDING SOME TERMS IN HAM RADIO 151

CHAPTER 17: TIPS FOR ENTHUSIASTS AS WELL AS HOBBYISTS 156

CHAPTER 18: HAM RADIO BASICS 159

CHAPTER 19: SELECTING A HAM RADIO 172

CONCLUSION 183

Introduction

Once you've got this book, you'll discover that the process of setting up an ham radio station is straightforward. If you're passionate about transmitting messages with an radio ham, you could start from scratch or make use of an existing radio kit.

I know how daunting this endeavor could be for novices to the field particularly since it's not a business that is purely for profit. This means that you aren't going into broadcasting amateur radio for profits.

The reason for starting the broadcasting of amateur radio is to broadcast messages and receive them during times of need. In these instances traditional media houses could have been shut down.

If you read this book, you'll be able to comprehend some of the phrases you might hear in the air. You'll know how to make it on the air, make contacts and get ragchewing with other radio amateurs. It allows you to interact on the air and connect with other hams from across the globe.

If you're interested in becoming an ham radio operator you should be enthusiastic about broadcasting and be able to browse the internet and find more information from knowledgeable hams.

Chapter 1: What's Ham Radio Station?

Ham radio stations are one of the radio stations designed to offer radio broadcasting services as part of the ham radio services intended for the ham radio user. Radio enthusiasts and hobbyists construct and operate various kinds of radio stations for ham radio such as mobile stations and temporary field stations ground stations, as well as space stations. Hams who broadcast radio are amateurs fascinated by radio services.

A "shack" is a slang term used to describe the location in which your radio station ham is situated. The name comes from the tiny space that is that is used to store batteries and radio equipment in the upper structures of a naval vessel.

Different types of Ham Radio Stations

Radio stations can be of different types. They include:

Mobile Ham Radio Stations

This kind of Ham radio station can be found on the vehicle. It comes with an antenna or two as well as a microphone and an transceiver. The transceiver is specifically designed to be mounted on vehicles. They tend to be smaller than transceivers specifically designed specifically for fixed or ground radio stations.

Mobile radios are designed to be mounted inside a trunk or beneath the seat. The radio comes with a detachable control unit that can be put in a different place in relation to the radio itself.

Ham Radio Portable or Temporary Station

Portable radio stations can be considered to be temporary since they are located in temporary locations. They are often constructed to broadcast messages of relief to areas affected by emergencies or in the event of a disaster. They also transmit public service messages or announcements at events, competitions or sports. This kind of station is also helpful in recreational activities as well as other outdoor events.

The equipment required to construct portable radio stations is similar to fixed or mobile radio stations. The main reason to choose it is the possibility of transferring batteries, transceivers antennas, batteries, and the power sources. This type of radio relies either on portable generators or batteries for source of energy. It is the best choice for communications over long distances.

Handheld Ham Radio Station

The radio station comes equipped with all the required equipment and gadgets that allow it to broadcast signals and communicate with other Hams. A handheld radio is equipped with an internal transceiver, an antenna, as well as a battery that is included in the box.

Handheld radios are specifically designed to work with the ultra-high frequency (UHF) or extremely high frequency (VHF) bands. In the majority of cases they are made for with Frequency Modulations (FM) broadcasts of voice only.

Repeater Ham Radio Station

The repeater radio station was intended to broaden and extend the scope of communication for other stations. It uses the receiver that is tuned to a specific radio frequency, and the transmitter is tuned to another frequency. It allows the transmitter to transmit using one frequency and then listen for signals at the second frequency.

You can create two-way communications by using a repeater ham radio station located on top of a structure, a tower or a mountain. This gives it an advantage over other stations in that all stations that aren't capable of interfacing with each other could use it as a broadcast in two directions to transmit and receive signals, and vice versa.

Fixed Ground, Permanent, or Fixed Ham Radio Station

It is an ham radio station that is built within a permanent structure and the equipment that is used for this kind of radio is not portable. Permanent, fixed, or ground radio stations are built in offices, schools as well as churches and homes, which includes public houses.

This kind of radio station is outfitted with an transceiver, one, or two antennas as well as microphones to transmit voice messages, Morse codes, and Telegraph keys. They can also be made for communication that use digital modes, such as PSK31 and the RTTY. It is also possible to build using a separate interface to connect transceivers with a computers with sound cards, but this isn't necessary to broadcast radio.

Other equipments used by fixed radio stations comprise SWR meters amplifiers, antenna rotators and many other accessories. The power source is usually an alternating current electrical source in the home. The other sources that power ground stations are batteries or generators for electricity.

Space Ham Radio Station

The space-ham radio station can be constructed in a spaceship or satellite, as well as an international satellite. Certain countries have rules regarding the operation that space stations operate under. A few of these stations may be located on satellites that orbit around the earth.

Space stations can function as transponders, repeaters or even transponders using automated control. They are often employed by permanent or ground radio stations that transmit their signals to station, either ground or fixed.

Despite the many radio stations, they all is identified with distinct call numbers. They are issued by authorities that regulate radio stations in different nations like the FCC.

How to Set up the Station

If you're interested in engaging in radio communications using ham radio it is necessary to establish the station yourself. The area where equipment used by ham radios is set up to operate is known as the radio shack. The designs of ham radio stations can be different based on personal preferences like:

Certain station designs could include just a transceiver in the edge of a room.

If you have additional equipment to your station You may require more space. It is possible to add some charts and wall maps.

Before you begin building your station, make sure you check the the available space and consider items you require to make it easy to transmit.

Take a look at the available space and think about the items that could be included in all the areas.

There are spare rooms you can think about in your house as well as attics, small cabinets or big ones, garages sheds in the garden and lofts. Each space you pick for your home has advantages and disadvantages.

Essential Equipment

The essential equipment needed to run an amateur radio shop includes

1. High-frequency transceivers are available in various types of HF equipment suitable for your station. They can be used to cover frequencies ranging from 1.8 milliseconds to 50 millimeters. This can extend to the lower parts in spectrum known as the Very High Frequency (VHF) spectrum.

2. VHF stands for Very High Frequency (VHF) transceivers are an electronic device that is handheld or mobile to broadcast FM radio.

3. Ultra-High Frequency (UHF) transceivers are an handheld or mobile device that transmits FM radio.

4. Auxiliary equipment for radio stations they are essential instruments to manage your radio station's operation, for example, VSWR meters for accessing the operation of your antenna, as well as the power sources to various systems. There are also antenna switches to control different antennas.

5. QRP equipment - designed for low power operation, allowing users to create the station of their choice. It uses Morse code to streamline the process. It also includes kits that will assist you in building your station.

Other things to consider prior to setting up your own radio station comprise

* The option of choosing between transmitters, receivers and other items of ancillary use.

* Mains power-provide enough space for it.

* Antenna feeders should allow for adequate access to.

* Make sure that the area isn't too cold or hot.

* Make sure you have security on the equipment, if it's located outside of your home.

* Ensure that the space is accessible for all times.

Let's now look at how you can setup your radio station on ham radio.

Install the proper lighting in the shack. Ensure that the the table on which the equipment is stored has enough lighting. Another source of illumination at the station is an angle lamps or a small strip lamp that can be placed under a shelf to illuminate the space.

The electrical wiring of your radio station. You need to plan the wiring for your radio station at the beginning of the construction. Plan out what your Mains electrical wiring as well as lighting will be. Connect the sockets in the mains wiring in order to provide to all the equipment needed. Install the multi-way connectors behind the table to make it easy to connect. Be cautious when wiring your radio shop and ensure that the circuits aren't overloaded. Make sure to install circuit breakers as well as cut-outs that are activated if there's an imbalance between the neutral line as well as the lines or an earth current are identified.

Another option is to refer to the regulations for wiring in your country to find recommendations.

Take into consideration the layout of your equipment The layout of your radio station's equipment is essential if you are planning to utilize the station for a prolonged period of time, especially during contests. Set your main transceiver or receiver on the center on the table. By using this arrangement, you effortlessly tune your radio while resting your hand upon the table. After that, you can be comfortable and comfortably transmit messages for extended intervals. A linear amplifier and a second receiver can be set up on the opposite end of the table.

Place the microphone on the left side of the table, so that you can always make use of your right hand to write or making notes. If you're making use of your Morse key, put it on the right-hand edge of your table as well. If you're a left-handed individual You can alter the places.

Take into consideration the safety of everyone at the Shack Safety of all employees who works within the shack is something to consider. Install RCCBs Residual Circuit Circuit Breakers and make your station child-proof so that injuries and accidents to children. Remove soldering irons from the station when you're not making use of them.

Before starting, you should engage in market surveys Go to the market to see the price and quality of the products you'd like to purchase. Other aspects to take into consideration are specifications for the product, power output and the overall performance that the device is operating at. Be sure to verify the dimensions and efficiency of the device prior to buying.

Think about the intended use of the transceiver prior buying It is the location in which you'll install it and the frequency you will use it. Examine the antennas you require as well as the type of interference, the dimensions of the station and the power source.

Things to consider before purchasing the High-Frequency Transceiver

A. Computer-linked control and linking the ability to control and link your transceiver via computers should be a top priority because software has been integrated into the ham radio communications.

b. Peak Power output: Consider an output of approximately 100 watts for your station. It is possible to use an amplifier that is linear to provide around 400 watts of power, as per the guidelines of some countries.

C. Take into consideration the expense of buying a new machine in comparison to a second-hand model which is functional.

D. Check the performance of the HF transceiver prior to purchasing.

e. Examine the dimensions of the transceiver or rig. In a small shack, using a transceiver of high performance is recommended. It is possible to have transceivers with large capacities however, medium-sized ones may be more efficient.

Think about the different types of transceivers available

The types of transceivers that you can purchase comprise

1. Base Unit transceivers. This includes a multiple-mode VHF, or VHF base station transmitter.

2. Mobile transceivers that are designed for use in mobile. Multi-mode transceivers are available available that could be used to create the base station.

3. Handheld transceivers: Purchase single handhelds , or dual or even triple handheld devices. It is based on the size of your shack and the expected performance.

Purchase Radio Linear Amplifier

This device is utilized to boost or enhance the power output of transceivers in which SSB which is an amplitude-sensitive signal, is transmitted. Linear amplifiers are favored when signals feature an amplitude component. They can be used to create the amplitude modulation technique, quadrature amplitude modulation and one sideband.

Linear Amplifiers

Two kinds of Linear Amplifiers comprise tube/valve ham radio linear amplifiers as well as the ham radio linear amplifiers made of semiconductors.

Things to consider when installing a Linear Amplifier

* Increasing output of the signal.

* Increased the power output for a better signal to the receiver.

* Make sure to check to see if there are local interfering signals.

Make sure antennas aren't installed on areas where there is no space.

• Improve your antenna as well as the feeder prior to buying a linear amplifier.

- Upgrade the elements such as feeders VSWR meters and antennas in order to deal with an increased power output.

* Take into consideration the cost.

Building Radio Kits by purchasing radio kits

Build kits that will assist you set up your radio station. It will increase your knowledge on the field. You'll also be able to manage radio frequency and electronic circuitry. In this instance you will need equipment for building the task.

Different types of Radio Kits There are a variety of radio kits available

1. Radio Frequency Amplifier kits to build audio frequency power amplifiers that can be used with low-power High-Frequency transmitters.

2. Low Power QRP transceivers and transmitters kits for Low Power QRP and High Frequency operation.

3. Kits for Simple Receiver - for ham radios in the on the HF band.

4. Kits for ancillary equipment For Morse training oscillators as well as small amplifiers timers, and a variety of other products.

Essential Items for Building Kits

The things that are required are

A) Test Meter - for example, a the basic multimeter used to locate the source of faults

B) Oscilloscope - for repairs at the station.

C) Soldering iron - used for soldering equipment parts.

D) basic tools - including screwdrivers and wire cutters, spanners and pliers.

Purchasing Ham Radio Antennas

There are many antennas that can be used that can be used by a ham radio station. A radio operator must be careful of which one is suitable in accordance with the dimension of the radio station.

Different kinds of Ham Radio Antennas

This includes

1. Wire Dipole antenna - one of the least expensive and easiest to utilize for transmitting ham radio. The higher up the antenna is placed upwards, the better performance.

2. The fed wire Antenna is a long-wire antenna that is simple to build and to install. It features the length of a wire connected to the transmitter or receiver. This is the most simple antenna to construct. An antenna tuning device is interconnected to the wire that feeds the antenna at its end and then placed between the transmitter or receiver as well as the wire for antenna.

3. Directional or Yagi Antennas that are ideal for High Frequency, Very High Frequency and Ultra High Frequency. The

Yagi antenna is used to transmit long distances.

4. The Inverted The V Dipole Antenna - functions with the same central high point on which it is positioned.

5. It is also known as the En-Fed Half Wave (EFHW) Antenna that is a half-wavelength in length and operates at the lowest frequency. It is fed by 500 coaxial cables that are compatible with an radio Frequency transformer, which is known as unun. It also includes the step-up impedance.

6. Vertical Antenna: the radiating is horizontal to earth's surface.

7. High-Frequency Wire Doublet antenna - this is a dipole that has an antenna tuning or matching unit, as well as an unrestricted wire feeder. It is designed for high-frequency bands.

What is the best way to set goals?

It is impossible for any organization to thrive without goals. These goals can help your ham radio station attain its desired goals. Find new and innovative methods of handling every situation.

Here are the best ways to set objectives that can guide your business:

1. Maintain integrity as a manager or an operator of a radio station operated by hams.

2. Delegate tasks to each member of your radio group. The way it's said by Stephen Covey "No participation, not a commitment."

3. Make a mission statement that your station's mission statement. share this mission statement with every participant. Make it a note and have everyone read it frequently.

4. Make a commitment to excellence and perform things to the best of your abilities.

5. You must take responsibility for your company.

6. Make sure you are enthusiastic about your projects and goals and bring excitement.

7. Keep a positive outlook throughout your life.

8. Develop relationships with your team members and keep the relationships.

9. Be respectful of people and take action.

10. Enjoy your wins Learn lessons from mistakes and enjoy yourself.

Making the Right Choices for Your Operation

The choice of your operation can aid in determining which radio model to make use of for your service. Before you get started with your operations you should know these four amateur radio profiles you should take into consideration. These profiles will meet your operational requirements

It is also advisable to talk with the local ham operators to find out which band is frequently used and easily within your region.

A few of the options you have to select from include:

An. Emcomm or Emergency Communication team member. It is often referred to by the name of personal emergency communication. This type of communication is designed to operate with easily accessible repeaters for small distances. You'll need the UHF and VHF dual-band which can be described as a hand-held radio having around fifty memories, 5 or 3 Watts of output power for this type of operation.

Other things you'll need are the possibility of a rechargeable battery backup and AAA or AA batteries, and an automotive adapter as well as a desktop quick charger.

Additionally, if you would like to operate with no repeaters or using remote repeaters in remote locations or to transmit via mobile include an VHF, UHF or VHF mobile radio that has the capacity of 25 watts.

B. Beginners or entry-level home high-frequency operation Use a high-frequency transceiver that has a 100 watt output as well as a built-in antenna tuner. It is possible to operate it directly with the alternating current source or external power supply with direct current. The device also features the multi-band dipole. It is a great and affordable antenna connected to radio via an open-wire feedline as well as coaxial cables. It may require an external tuner for the antenna.

c. Local or regional Operation - Use a UHF or VHF mobile radio that has an output power of 25 watts and higher. You can then use an asymmetrical quarter-wave dual-band mobile whip antenna. However, if you're broadcasting at home, it is possible to utilize an indoor mobile antenna, or mount an outdoor base antenna with dual bands. Attach the radio and the antenna by using an coaxial cable.

D. The entire band (either portable or mobile) Operation - Use an all-band and all-mode portable transceiver that covers the High Frequency and Very high-frequency bands. It also includes bands of 70 centimeters in certain modes.

How do you allocate your resources

If you are a novice in amateur radio you must allocate your resources efficiently. This will enable you make the right choices for your ham radio shack including the kind of antenna, the types of broadcasts, and the kind of radio.

Here are the best ways to distribute your resources:

1. Create a plan for your resources , in terms of required equipment, setup and select the kind of ham radio that you will

purchase. As I mentioned earlier, make a plan for the space you have available within your ham shack in order to ensure ease of use during operation.

2. Select a reliable radio to run your ham operation This will require an enormous amount of your budget. Choose Shortwave or High Frequency radios that are capable of operating under 30 MHz.

3. Find the right radio accessories, like microphones, but certain radios have microphones. It is possible to purchase headphones that allow you to pair with a boom mic. Certain microphones can withstand high frequencies and perform well in these situations.

4. Purchase the best antennas and computers. Putting the correct antenna and computer in your shack will enhance the efficiency of your operation greatly. The antenna is the one that determines the strength of your radio frequency , and

also enhances your radio transmissions. Select the appropriate antenna support.

What Radio Should You Choose for Operation

The selection of a radio device that is suitable for broadcasting on ham radio is essential to judge the quality of your operations. Unwise choices could result in poor communications.

If you are looking to make the right decision be sure to consider the specifications of the gadget, such as the range of VHF, UHF and HF, and the power. Other features include backlit display as well as frequency range, memory and bands. It will let you determine if it's the right option for you.

1. Yaesu FT-70DR Ham Radio

It is widely regarded as the most suitable radio for novices. It comes with a huge screen for crisp and clear watching of channels when you turn the knobs, or whenever you alter the settings.

The two dials on the top determine the frequency and volume. Yaesu features an auto gain control that enhances or diminishes your voice according to the strength of your signal.

The features of Yaesu

* It is equipped with a light keypad.

* A 70 cm amateur radio.

* 5 watts of stable Power output for RF.

Display of Alphanumeric characters.

* Yaesu allows you to sort through up to a thousand channels.

* Designed using an 1400mAh battery set.

2. Baofeng UV5RA Ham Two Way Radio

Baofeng radio is among the cheapest handheld radios you can utilize for transmitting. It's designed using UHF as well as VHF support. The keys are double-pTT, as well as unadjustable power levels. This helps to conserve the battery's power. It features a LED flashlight and an FM radio reception for commercial use. There is a huge LCD screen to allow you to read anytime.

The main feature on Baofeng Two-Way radio

* It comes with two PTT keys.

* Doable for CTCSS or DCS scanning.

* Created with an emergency alarm feature as well as a keypad lock.

4. 4 Watts power consumption.

*UHF, as well as VHF support.

* Background with light colors that you can pick from.

3. MD-380 from TYT Ham Radio

The TYT radio comes with 16 memory channels, but it isn't as good in the event that you wish access international radio stations. The UHF model has two different types of antennas including the stubby as well as broadband. The whip is however, equipped with superior performance over the stubby version.

The radio features of TYT

* Functions in one or Two mode.

* Created as analog and digital , with the belt clip as well as AC charging

Helps you program software and program cables.

* It comes with a UK Three-pin power supply.

* Works with 16 channel channels. It also runs in tier 1 and 2 modes.

4. Kenwood Original TH-D74A Ham Radio

This is a handheld handheld TH-D74A radio with a trans-reflective display. It's a tri-band radio that is well-built and features an easy-to-read display. It is constructed with the highest quality materials. it is Kenwood radio has weatherproofing, and has an IP54/55 rating. This indicates that it's waterproof and dustproof. It is a great mode that has the lowest step frequency range of around 20 HZ SSB.

Kenwood's Features Kenwood

* It has APRS and KISS, both with 5 watts of power output.

* It includes an external decode feature.

* Kenwood has a pitch frequency

* The radio supports DStar digital protocol.

* Built using four TX Power selections and GPS.

The system is capable of taking AM station CB stations, and FB stations. stations with a height of around 11 meters.

It is tri-band Ham radio with two frequencies: 144 MHz and 220 MHz/430 (MHz).

Radio Accessories

A few of the radio equipment you'll require for your ham radio operation include

1. Base station radios - headphones come with hand-held microphones. Also, you can use boomsets made up of desk microphones from third parties and headset booms. These sets can boost in the sound quality and clarity of broadcasts. You will be able to hear other stations with greater clarity by using a headset in chaotic or emergency situation.

2. Battery chargers have an extra battery that allows you to switch batteries when the battery is depleted. Make use of a quick charger in order to speed up charging.

3. Front panels detachable - many radios for hams operate without the front panels being disconnected of the system. You'll need an additional control cable to control these panels.

4. Smart microphones are handheld microphones that can be used for mobile stations or base stations. They come with small keys and buttons for front panels.

5. The ability to tune your antenna using an Radio Frequency Meter and an SWR bridge can help regulate the power flow into and out of your antenna.

With Software-Enabled Radios

Certain transceivers for ham radios come with built-in computers which are designed with software or firmware that functions as transceivers. The software is from the first manufacturers of these devices.

Other appropriate software is required to manage the transceiver or receiver, if it is not controlled by the front panel. This includes Kenwood TS-B2000 and Ten-Tec Pegasus. Kenwood TS-B2000 and Ten-Tec Pegasus. These transceivers come with PC software that provides an interface for human-computer interaction.

Radios Suitable for High-Frequency Bands

As per the International Telecommunications Union (ITU) the High-Frequency bands were designed to accommodate radio waves that range between 3 MHz to 30 MHz. This is known as decameter also known as decameter bands. In this case, wavelengths vary between one and 10 decameters.

A radio designed for band with high frequency is called that of the Citizens Band (CB) radio.

Radios Suitable for Ultra High-Frequency Bands

Additionally also, the ITU designates Ultra High-Frequency (UHF) antennas for radio frequencies in 300 MHz and 3GHz. The wavelength range extends from 1 meter to 1/10 meter, and is referred to in the decimeter range.

Beyond this range there are radio bands referred to as Super-High Frequency (SHF) which are are thought of as microwave frequencies. They are generally propagated through line of sight , but they are also is blocked by mountains and huge structures. Indoor transmissions are when these signals can penetrate through the walls of buildings.

Radios that work on UHF bands include hand-held radios, walkie talkies, two-way mobile systems for land two-way radios inside cell phones, vehicles and portable wireless devices and cordless phones, such as Motorola RMU2040 UHF radio as well as Motorola RDU4100 4-watt UHF radio.

Radios Suitable for Very High-Frequency Bands

VHF radios work between 136 MHz and 174 the MHz. VHF radio waves are longer and more close to their ground than the other wave. Thus, it is able to cover greater distances while using less power. If there isn't any disturbance, the two-way radios work more efficiently with clear lines of sight between the transmitter and the receiver.

These radios are utilized for marine and aviation broadcasts in which signals are transmitted directly from space to earth or over open bodies of water with minimal or obstructions.

Some examples of these radios include walkie-talkies, such as the Motorola RMV2080 VHF Radio. It's got two-watts and eight channels. Another is the Motorola RDV5100 VHF Radio that has five-watts and 10 channels.

What are the reasons for noise and filtering?

Noise can be heard between high frequencies between 3 Mhz and 30 MHz. In this frequency range the antenna will detect noise, no matter the location it is placed. This is considered to be radio background noise. The normal noise level will decrease the frequency of your radio if it rises to the area of the Very high

Frequency (VHF) over 100 MHz. This is the thermal noise of the front-end.

Filtering refers to the process of making use of an electric circuit known as a network that allows frequencies to pass , but blocks unwanted frequencies or reduces. There are four kinds of filters, including Bandstop, HIghpass, Lowpass as well as Bandpass (Notch).

Chapter 2: What To Pick Antennas For Ham Radio

Noise can come from any direction but the desired sound is only heard in only one direction. An excellent antenna can help make a weak radio sound more appealing. Thus it is recommended to use an antenna.

For operation, the operator are able to utilize 100 feet of RG-58 and an 66-foot dipole which can be tuned to all frequencies. However, you could be losing more than half the output of the transmitter on the bands that have higher frequency. There will be signals transmitted through your coaxial cable. If your signals are weak or the frequency is too noisy you'll lose contacts. It is possible to use an RF transformer as well as an impedance step-up to increase the frequency.

Types of antennas that you can pick from are:

1. Dipole antenna

2. Microwave antenna

3. Log periodic antenna

4. Reflector antenna

5. Antennas Dual Band UHF/VHF

6. Wire antenna

7. EFHW antenna

8. Antenna for a traveling wave.

Using High-Frequency Antennas

It is possible to use high-frequency antennas like wire antennas, also known as wire dipoles as well as the rhombic antennas in higher frequencies. Other antennas include quads or log-periodic, as well as yagi.

If you are looking to receive higher-frequency signals you should utilize an antenna made of random wire. It's the same with an antenna with directional characteristics that can be useful to transmit signals.

Using Ultra High-Frequency Antennas

Ultra High-Frequency antennas were designed to operate in the radio frequency 300 MHz to 3 GHz. It is also known as the decimeter band since the the wavelength extends from 1 meter to 1/10 meter.

But, the radio bands that are beyond this range are referred in the term Super High Frequency (SHF). They are also considered to be a radio frequency bands that are microwave. Radio signals from UHF are transmitted via line of sight, however they are typically blocked by tall hills and buildings. Indoor signals have enough power to go through walls.

UHF antennas tend to be small. For instance, the most common multidirectional antenna, the quarter-wavemonopole, can be between 2.5 centimeters to 25 centimeters. Ultra High-Frequency wavelengths are so short , making the antennas tiny enough to be used on handheld or mobile radios.

Some examples of these antennas include Omnidirectional antennas, also known as short whips as well as rubber ducky antennas. dipoles in sleeves as well as an (PIFA) planar Inverted F antenna, which is commonly used on mobile phones.

Utilizing very high-frequency antennas

They are longer than the ones that are built into Ultra-High-Frequency radios. They can be useful in boosting transmission over longer distances , and are ideal for very high-frequency ranges. They can also are able to penetrate the foliage and trees more then UHF models.

Management of Connectors and Feedline

Radios can be connected to antennas by feeding lines. Different feedlines are available in the marketplace, however you can utilize coaxial cables to connect your radio station ham in lieu of feedlines.

Coaxial cables are able to be installed quickly. This can be seen on Radio Frequency signals occurring between an antenna and a radio. The loss is increased when frequencies of the signals that are transmitted via the coaxial cable is also increased.

To manage a feedline make sure that the impedance of the feeder to both the impedance at which it is input by your antenna and its output impedance. Impedance is a measure of resistance to the flow of alternating current within the circuit. The measurement unit for the impedance of a circuit is Ohms.

The transmitters of the majority of Ham radios are made to operate at an impedance of 50 Ohms. This is why impedance is the most simple coaxial cable to use in the in a ham radio shack.

One of the most common connectors that are used in coaxial cables for Ham radio stations is the PL-259 connector. It's also used for high Frequencies but isn't the ideal connector for high frequency. A connector that is suitable for high Frequency can be found in that of the Type N connector that can be able to withstand frequencies of over 400 MHz.

If you are looking to stop loss of feedline, make sure that you be sure to tighten the connectors thoroughly and seal them with sealant to keep water out of the system. Soldering connectors and tightening antenna connections are the best ways to stop feedline loss.

How do you support your antennas

Antennas require support to ensure the proper transmitting signals. This can prevent damage caused by hurricanes, wind or hurricanes. There are a variety of structures that can help support the antenna, such as:

Tubular steel structures made of steel - built to safeguard the cables as well as other structure in.

Reinforced concrete towers - these kinds of towers made of concrete are costly to construct, but are useful for helping to support antennas.

Lattice made of steel - this is a typical construction used for antennas that support them, as it is durable resilient to weather and wind and lightweight.

Drones, drones or unmanned aerial vehicles may be equipped to serve as a support to antennas to enhance radio communications.

Wooden towers serve as supports for antennas.

Telescopic mast is a kind of radio tower that can be constructed quickly to provide an antenna support.

Materials made of composites and fiberglass are ideal for supporting antennas to support medium-wave transmitters, or beacons with low power.

Mast radiators are antenna supports designed to be radio towers to support medium or long wave transmissions.

Poles These are made of guyed wood as well as self-supporting wood structures. Galvanized steel material could be used for poles.

Tall buildings are used to support antennas.

Kites and Balloons : Kites or balloons made from leather can be used as supports for antennas.

Care for your Antennas and trees

The tallest structures, high buildings and trees may block the signals of your antennas. If you have tall trees like the pines and elms that surround your home, they may influence your signal. There are two kinds of television antennas like the exterior and indoor antennas. Both antennas should be able of detecting VHF as well as UHF channels. This is the requirement of the FCC, Federal Communications Commission.

However, trees that are tall or nearby bushes could block radio frequencies, specifically in the case of indoor antennas. Leaves of such trees may reflect the signals from your station . During the dry weather, they transmit more signal. If this happens you'll notice shaking, jerky, or frozen images in your phone.

Utilizing Tripods and Masts

Masts and towers are large structures that are used to support antennas as well as television, broadcasting and telecoms. Masts of two kinds are self-supporting structures as well as guyed structures. The mast can be described as a ground- or roof-top structure designed to support antennas. They are situated at high elevations which allow them to send as well as receive information from other contacts.

If you are able to rotate your antenna, you'll be able stop interference from one direction or increase the signal in another direction. To accomplish this you need to place a small high-frequency beam or antenna mounted on a tripod, on the top of a mast or roof.

It is recommended to make use of a rotator that is mounted to the supporting and rotates the antenna in different directions. You can operate the rotator directly at your desk within the station, and track the direction with the gauge.

Controlling Towers and Rotators

Towers and Rotators

Antennas are designed so that they face different directions based on the signal expected. Certain ham operators employ small antennas, or larger ones, though they might not require as the same amount of power as other communication services, but by using antennas, they can generate the power of radiation.

To be able to focus in the right direction, the antenna requires the help of a rotator, particularly when it is placed on an elevated tower. There are antennas that can be controlled and alter their phases by switching them on or off. The device that lets you alter the phase of your radio antenna is known as the Rotator.

Things to take into consideration when working with rotators include the size and wind load, the wiring control mechanisms, turning torque, the ability to brake as well as the vertical load and direction.

The Sizes Antenna Rotators Antenna Rotators are smaller size, but they are also portable and work with 12 VDC.

Wind Load - The largest antenna area as square feet, is called the wind load.

Wiring - figure out what wiring to run to the controller as well as the Rotator. It is the Green Heron has provided a wireless solution for this, however it is necessary to connect power to the rotator of the tower prior to it being able to function.

Controls - Each rotator is equipped with customized control units that regulate the functions within the system. This can also assist in displaying the locations or the phases of an antenna. Certain rotators are equipped with software that allows to make it easy to control, such as the Easy-Rotor Controller or the Rotor-EZ.

Turning Torque, or Braking Ability The maximum turn or twist the device can endure is its braking capacity. Turning torque is the highest amount of torque a rotational rotator produces to rotate the antennas.

Vertical Load: The maximum vertical load is shown in pounds and kilograms. In the event that the antenna's weight is larger in weight, it will be much more challenging to move.

Direction - establish your direction by using the topo map and Compass. These are great for actualizing meaningful signals but if you are targeting long-distance contacts, you need to use azimuthally-equidistant maps.

Chapter 3: What To Pick The Right Computer To Run The Ham Radio Station Of Your Choice. Ham Radio Station

Amateur radio stations have taken on the computer. The device was originally utilized in ham shacks substitute for log books. Today, it's an additional device that is employed to manage the operation that include transmitting and receiving CW and connecting your station with other Hams online.

How do you choose your computer? It's a windows-based system that includes additional radio applications running in the system operating the computer. It is possible to choose between the Windows operating system, or a Mac operating system. Both computers are reliable and reliable for radio operation.

Another useful device for amateur radio operation can be the Mac. If you are interested in using an Macintosh system to operate your station, join the Macintosh here.

Utilizing using a Mac as well as a Personal Computer (PC)

Mac computing systems are getting more ground on the radio industry. The community of amateur radio enthusiasts has been enjoying incredible information via Ham-Mac's mailing lists. Ham-Mac.

You can also opt to utilize Windows computers to manage your ham radio operation. Whatever system you choose to choose, different tools, programs and programs can assist to manage the radio station. These tools will allow you manage your broadcasts in the way that you'd like.

There are various commercially produced software available from ham radio community and clubs through sharing their information online.

What is Radio Controls?

The radio controls (RCs) within an ham radio use using the USB control interface or RS-232 which you can utilize to control and monitor each radio function. It is a method of operating a device with remotes that work with radio signals that are transmitted from the distance.

If you push the button, electrical signals or electrical impulses are transmitted through the transmitter to the air for a response from the device that is in response which could be the entrance to garages or other devices.

Digital Modes

Digital modes are now accessable via an audio card. It's the standard device to transmit and receive data. You can transform your computer and radio into an incredible data center with an easy device that controls data and radio.

Some of the most popular companies that make data interfaces include West Mountain Radio and MFJ Enterprises. You can choose to utilize the Kantronics KAM, the MFJ-1278B Timewave PK-232, or DSP-599zx. These are all external controllers for multiple modes that operate in digital modes. The only application you'll need to operate these modes is Hyperterm which is a terminal application that is built into Windows.

Hardware Considerations for Your Station

If you decide to add computers into your Ham shack, then you don't require the latest high-speed computer. A dated device that's been in your home for a long time can perform similar work.

It will allow you to run complex computer applications like high-performance data modes, antenna modeling and high-performance data modems. Computers can assist you to record data and monitoring the APRS website.

If you have new computers for your radio station it is recommended to include the serial port expansion card or use a converter USB-to-RS-232. There aren't any problems like compatibility issues or driver issues with the serial port expansion cards you have. But, the USB converters are much easier to set up on devices. Today, it's easier to convert additional accessories or radios into USB interfaces.

Controlling an Remote Control Radio Station

A lot of hams use remote control devices for managing their stations. This is the case if you are unable to make use of an efficient antenna in the area in the area where you reside or where you've set the radio station. In the majority of cases, the landlord might not permit such installation. However, without an antenna to run an amateur radio channel, it, it is challenging.

Another reason why the owner might not allow installation of antennas may be because of noise. Your antennas can cause noise in certain electronic devices, power lines computers, and other electrical appliances.

To prevent interference To avoid interferences, you can put up your radio station as well as its antenna in a different location and use a remote to operate it. Additionally, you could make use of the internet to manage the station. This is similar to using a microphone that has the longest cord available to control an internet radio station from your residence.

Remote Control Guideline for the use of a Ham Radio

Before you begin setting your amateur radio station Here are some rules to follow:

a. Always keep your identity in mind using your name and call sign. It is a way to indicate your current location.

b. Manage and control your radio's radio stations by shutting off your radio transmitter. You can utilize some security control devices that are accessible via telephone to turn off or the relay responsible for supplying AC power for your station's shack.

C. You should seek permission to operate the radio transmitter at your area, regardless of whether you are licensed or not.

d. You must obtain operating permits (a license) that allows you to establish the transmitter at your place. It can be an individual local licence or regional license, or any other similar licenses from regulatory bodies that allow you to broadcast messages on the radio.

How do you access a Remote Control Station

It is possible to access the remote control station via computers and an internet connection at each of the locations. There is also a DTMF controller attached with the telephone line so that it doesn't lose control over the radio. After that, you'll need to shut down the entire system and begin again.

In certain situations you will need to install the station at home , and be able to control it outside of home such as in a hotel room, or in other locations. You can install the control software on your laptop and control the station from outside. station.

Another alternative is that you don't need to carry your laptop around; your smartphone or tablet can be enough to control the station by using programs like Pignology. This program works on iPhones as well as other tablets for easy control. This technology can create or share radio stations with certain clubs or associations.

How to Purchase Used or new equipment

Ham radio equipment is available to purchase in the marketplace. You can buy second-hand or new devices based on the resources you have and the space that you will need for your station. It is suggested to purchase new equipment however it can be more expensive than previously owned equipment. It is important to be careful when buying used devices to avoid buying the device that has issues.

Here are some ways to choose the right equipment, whether new or used, for your ham shack.

1. Find out what kind of equipment you will require. Include the bands that will be covered by your transmitter, the types of operation, such as FM AM, digimodes and SSB. Also, you should decide on the need for power output and filtering.

2. Choose if you're in need of a new machine or an old one.

3. Find out if you'll go through a dealer or a private sale.

Things to examine on previously owned equipment prior to purchasing

a. Examine all buttons on your device and confirm they're functioning properly.

B. Examine the oscillator's condition for drifts the used equipment.

C. Check the design of the equipment to make sure that it's neat and well maintained.

How to upgrade your Amateur Radio Station

It is possible to upgrade your license when you hold the General license and wish to upgrade to an additional class. It can be done by passing a 50-question multiple-choice test. The test will cover radio equipment design as well as advanced electronic theory. No Morse code questions involved.

If you hold an additional class license If you have an extra class license, the FCC allows you to use all frequencies assigned for the service Amateur.

Chapter 4: How To Organize The Operation Of Your Ham Radio Station

The organization of your radio station will enable you to enjoy an enjoyable time broadcasting from your ham radio station. There are a variety of methods to arrange your station for the most enjoyable radio experience on air:

* Choose and design an exciting design for your radio station.

Make sure that the electrical safety is maintained throughout and after the operation at the station.

* Control the radio frequency effects.

* Log in and verify your contact details.

Ground and bond the radio device.

How to Create the perfect Amateur Radio Station

To plan your ham radio station think about how you will map out the layout of your station, including the various rounds of sketches. Also, it requires human engineering in order to arrive at the ultimate arrangement for the space.

Also, plan your storage space, including the support system for your radio equipment.

Here are the steps to design your station:

1. Design and construction of radio stations.

2. Design and installation of an antenna system.

3. The construction of the tower is to support antennas.

4. The integration of the radio station

How to Make Use of the Station Notebook

A station notebook is essential to keep your notes as well as other data in one location. Every page includes a table as well as space to write down particulars such as Frequency Modus, QTH callsign Report received or sent UTC ON or OFF and dates.

How to view some examples

A good example from Ham Radio Station

Space Station

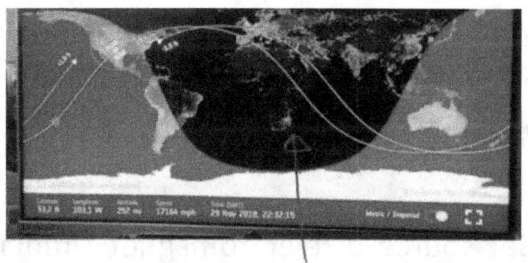

Image taken from an orbiting space station

Monitoring radio Frequency and electrical safety

The radio frequencies get absorbed into human tissues, but the effect is the sensation of heat. The RF frequencies can heat up the body's systems. The amount of absorption from RF frequencies is dependent on how much frequency the signal is. The maximum limit for total body tissue exposure lies in the region of 30 MHz to 300 MHz. In this situation our body's tissues absorb radiofrequency best when the whole human body has been exposed radio frequency.

If you're interested in ensuring your safety from the harmful effects of radio frequency radiation Here are some actions to follow:

* Set source of electromagnetic radiation away.

Make sure your bedroom is free of RFs.

* Make sure that you use your smartphone in a safe way of your mobile.

* Replace wireless devices with wired devices.

* Turn off wireless functions like routers, printers as well as other wireless devices.

How can you improve electrical Safety

Make sure you follow security practices for your ham radio station as well as your antenna. Avoid contact with the conductors of powerlines. Be aware when installing of antennas. This is due to the fact that antennas are powered by electricity mains. Be aware when opening the radio equipment to perform servicing or repairs.

Handling RF Exposures

If you're an owner of a mobile device The safe limit for mobile phones can be set at 1.6 milliwatts for each kg(1.6W/kg). This is averaged over one Gram of tissue. But, the compliance with the limit has to be proven prior to FCC approves for the marketing of phones within the United States.

So, in order to reduce exposure to radio frequencies you must reduce the time you are using the mobile phone. Utilize earbuds and headphones to maintain a space between you and your phone. If your signal is not strong, refrain from making calls. This could make the phone boost the RF power.

How do I apply First Aid at the station

First aid is a must method of handling emergencies at an Ham radio station because of unpredictability of the situation. Here are some tips to consider when you are an Amateur Radio Emergency services member.

a. Be sure that your family as well as team members remain secure.

b. Verify the security of your home before responding as volunteer.

c. Monitor your local ARES emergency net frequency.

D. Follow the directions that you were given by the ARES official.

It is important to note that the emergency coordinators are. Contact your local emergency coordinator for additional instructions.

Controlling the Effects of Lighting

Lightning can cause interference to your radio signals. This is a regular occurrence in our everyday lives especially in urban areas, accompanied by a technological revolution in wireless.

Signs of interference can be seen when radio frequency energy is transferred to an active device IC or tube inside the equipment or transistor. The energy is a source of conduction or radiation. When it travels throughout the atmosphere, the internal wiring acts as an antenna, which absorbs signals before transferring them into an active device.

Here are some ways to handle the signals:

a. Locate and locate the source of the interference, which is caused by broadband radiated and conducted RFI by using an AM radio tuned to an unheard frequency.

b. Install RFI filters on the offending signal path that causes the interference.

C. Take into consideration routing and ensure that cables are cut to the maximum extent possible.

d. Be sure to have a good connection because connectors left in place for a prolonged period of time may turn into metal oxide detectors to detect the radio frequency (RF).

e. Beware of connecting grounds that are not needed because this may increase the ground noise circulating.

F. Apply ground isolation devices in signal paths that are difficult to follow.

Direct Current and Alternating Current

If you wish that your equipment for radio to function, it has to be connected to an source of electricity. Different kinds of current are required for different components of a radio set-up. While electrons are all similar, the rate of flow is different. There are two kinds of electric currents that are alternating and direct current.

Alternating Current

The simplest description of an oscillating current would be current which flows both forward and backwards. This is what happens in the electrical sockets of your home.

Direct Current

This is the flow of current only in only one direction. Direct current is normally produced by a battery. For example, a resistor connected to the battery. In this instance the current is flowing from the positive terminal through the resistor and into that battery's negative side. The current in this instance is not affected by the length of time.

How to Control Radio Frequencies

Radio frequencies are present throughout our lives, from cellphones, to the airplanes, the internet and even when we look at weather forecasts. The management of RF helps minimize the effects of electromagnetic waves on human.

Bonding and Grounding

In the case of a ham radio station bonding and grounding are essential steps to protect your shack from the damaging effects from lightning strikes, alternating current as well as radio frequency. In the event of lightning the entire equipment is placed in the exact voltage. The function disperses the energy of the lightning on the Earth and then redirects it away from devices.

It also promotes AC safety and shields against the dangers of shock from equipment powered by alternating current. It offers a safe pathway for all equipment in the event that the wiring is damaged or insulation occurs.

Another benefit of bonding and grounding can stop the effects of radio frequencies from disrupting normal functioning of the radio equipment.

Bonding and grounding allows you to comply with the rules of lightning protection and communication systems, such as those in the National Electrical Code.

Chapter 5: What To Do Keep The Log Of Your Contacts

The need to keep track for your contact information is vital because of various reasons like:

a. Logs for operational factors that assist you in filling in your DX QSL card you have collected.

b. Personal aspects - logs allow you to keep track of your interactions with, as well as contests you participated in as well as awards you won.

C. Legal considerations - a record of your communications will provide evidence if there's a report of interference in your transmissions.

Here are some methods to keep the log on your connections:

1. Input the frequency, the date power output, as well as the mode of your operation.

2. Note the place, the name and date the contact was initiated and ended. Also, note their phone signal and report on their signal.

How to Use Computers and Log

There are two kinds of logs that are available at every station like:

1. Details about your radio station.

2. Operational details.

For a computer to enter the data it is necessary to install certain software. These programs can be set up to automatically update your information. It is also possible to make use of rotor control in order to create circles from your location to any location across the globe. You can also export your log book to the LOTW (Logbook of the World) by using the Export feature in the system.

How to submit an Entry Log for a Contest

You can upload the contest log by clicking on a method to submit. You can send the logs in the format that you prefer. You can send them directly to email addresses for the contest only to be used for straight key nights. Make sure you don't compress the files before sending the document as an attachment in your e-mail.

If you wish to submit your entries online, use the Web application. Visit the page you received of your Logs and view all the files submitted to contest. The files are automatically updated multiple times throughout the day.

What exactly are QSL Cards?

This is a document that confirms the existence of two-way radio communication which takes place between two radio stations. This also includes the reception of a single-way signal using an AM radio, as well as FM radio, as well as television.

A QSL card is similar to an ordinary post card. It is the same dimensions and is made of the same material. QSL is a QSL originates in the form of an International Q code meaning "I confirm receipt of your message" QSL cards include vintage QSL gallery, QSL managers, QSL cards, QSL bureau, and electronic QSLs.

How to send and receive OSLs

Ham radio stations can be able to send or receive QSL cards that have an addressed envelope directly to the station's manager. It can also be delivered to the station's own office to those who have a different person responsible for managing their cards. So, the envelope should contain your personal information including your address as well as the country you reside in.

How do I QSL Electronically

Electronic QSLs are favored because of the low cost of the card and low delivery costs.

Additionally, it offers more functionality, automation, and faster speeds.

Paper QSLs can be delayed when delivered when compared with electronic card. You can get electronic QSLs quicker and more easily. One of the challenges that this format has is forgery. To stop forgery, QSLs have been designed with security built-in. This makes them very difficult to forge.

Direct QSLs

You can do a direct QSL to a QSL manager. In this scenario the return postage is included in the cards. It's sent in the same way as it is an IRC (International Reply Coupon). It is exchanged in exchange for surface mail return postage , and the QSL managers will ask for the payment for airmail and other expenses.

The number of required International Reply Coupons listed in the bulletin. If the IRCs aren't included the cards will never be returned. In the majority of cases, it is possible to return it through the bureau.

How to Make Use of QSL Managers

In countries with electronic communication and a reliable postal system QSL managers can deliver cards to DXpeditions as well as stations located in remote areas. If you're listening to the bands, unusual DX stations might proclaim that they have an QSL manager that receives cards.

The QSL manager manages QSL activities on behalf of DX station.

Here are a few reasons you might require the help of a QSL manager:

1. When the operation on the radio is temporary, it's a temporary one.

2. If there isn't a QSL bureau in the country, the government can arrange for a radio company to be managed by a.

3. Make the most of the time you have available.

4. Stations are often located in areas inaccessible to postage service.

Utilizing Bureau or QSL Services

You can make use of bureau to transfer cards to an QSL manager. Ensure that the number of the manager is clearly highlighted within the card. This might be Station 1 or station 2. This way the card could be routed to the QSL the manager's location.

Application for awards: How do I apply?

If you are an amateur radio operator you are eligible to apply for awards through joining ARRL. It's an online community where amateur radio operators meet and discuss problems with the ARRL leadership.

You can also receive awards as a radio operator when you create a two-way broadcasting system that is shared with other stations. This is also known as contesting. It is a technique used to identify who can be contacted by other radio stations in certain bands, following a set rules, and within a certain period of time.

These awards require you to present a proof, for example QSL cards of your contacts that are in compliance with the awards.

Chapter 6: How To Operate Away From Home

To be able to operate from a different location than your home, create a remote-controlled station. This allows you to operate the station from the base of operation. Additionally, a third party may be able to manage the operation for you. The station may function by using an automated control.

Another method to control your radio station while outside of your home is using an control link. Through this link, you can utilize devices like a smartphone or laptop to manage your radio station when you're in your hotel, on the beach, or at other recreation spots.

Mobile Phones that Operate

Radio amateurs have learned how to put mobile stations inside their vehicles and then use them. The stations work properly with antennas designed for higher performance. It is possible to install low-quality antennas, and still be able to make contact using radio shacks that are located hundreds of miles away. Another option is mobile antennas.

If you're unable to utilize antennas in your home due to restrictions set by your landlord, then mobile radio stations will allow you locate better places to manage your business.

Using HF Mobile Radios

For high-frequency handheld radios the frequencies must be within 3 MHz and 30 MHz. The signal is transmitted by either the unit that is base or handheld transceiver that is broadcast into the atmosphere.

The radio waves return to earth after bounces off the Ionosphere. A transceiver that is tuned to the similar radio channel will also receive these radio waves.

Connecting Mobile Antennas

Select the correct type of mobile antenna that you want to put on the radio stations you are using. The way you set up your antenna is what determines its performance. But, you can improve the efficiency of the antenna by implementing a few guidelines and rules.

These are some of the general guidelines to keep in mind when it comes to antennas for mobile devices:

a. Choose a place that the antenna will be able to turn around.

b. Nearby objects could hinder the operation that your antenna is able to perform.

C. Find the best places to secure antennas.

D. You might consider the possibility of using an antenna for your radio inside.

Handling portable operations

The term "portable" refers to the station operates outside of its initial address. This is typically the licensed or registered address. In transmissions, radio operators from ham radio are required to add '/P' to the call number.

Using Portable Antennas

Small antennas that are able to be transported from one location to another is a portable antenna. These kinds of antennas can make the work of a ham a lot easier. It is possible to install portable antennas as permanent or temporary ones.

Connecting Portable Power Sources to the Internet

If you require an alternative source of power to power your ham radio you can buy an electric battery that has an maximum voltage that ranges from 9 volts and 15 Volts. Additionally, you can purchase an 12 volt battery regardless of the kind. However, you can make them 100 percent efficient with accessories.

They can comprise commercial AC or AC via an inverter, connected to a battery. They also could be a the battery in a vehicle.

What exactly is Field Day?

Field day refers to the overtures which take place when military is in operation. The term originated from military exercises. However, by the 1800s, the term was used to mean events in the field, such as hunts, picnics, dances and more.

In the field of amateur radio broadcasting it is about social outreach, technical abilities public service, as well as emergency preparedness for an event. For ham radio broadcasting field day is a cherished annual event that was first introduced in 1933.

Chapter 7: Understanding Hands-On Radio

Ham radio utilizes radio frequency signals for international and local broadcasts. It's a non-monetary exchange of information, sports contests, wireless experiments, and emergency communications.

The owner of this device is an amateur who is an enthusiast of radio or a hobbyist. This is because the International Telecommunications Union authorized this service by regulating radio. The operator must however be licensed prior to the ability to use the service.

How do you acquire tools and other equipment

The tools and equipment required to transmit messages via radio frequencies is the transceiver as well as the radio. The transceiver is able to be mobile, mounted, and handheld.

In addition, you can choose to add weather stations, computers towers, power cables as well as scanners, receivers, and scanners.

Utilizing tools for maintenance

The tools that you will need to maintain your station make use of in your radio amateur station are pliers, screws wire cutters, pliers wrenches, tool boxes, tool boxes and socket sets tape measure, electrical tape as well as an adjustable benches vice (permanently mounted or temporarily) as well as a utility knife ladder as well as a soldering iron.

Repairing and building tools

A few of the tools for repair and construction you'll need for your radio station are coax seals, sealers, additional connectors for coax ends, pencils and pad for notes, fuses and additional solder.

Accessories for Building and Restoring an Station

A few of the equipments that can be used to construct and repair the radio station include an antenna analyzer, reliable standby receivers digital multimeterswith dummy loads along with power, or SWR meters.

Other accessories to add to your radio station are external speakers, headphones, as well as lighting fixtures for your bench.

How to Keep Your Station Running

You can keep your radio's ham radio's shack checking your amplifiers and transmitters to ensure clear and crisp audio. Examine your radio frequency cable as well as antennas. This includes power output.

Make sure you verify the level of signals on all bands to determine if the levels are low enough or high. Check each antenna and evaluate their Standard Wave Ratios on all antennas. Check all feedlines outside including antennas using the aid of binoculars.

Examine the guy wires and ropes for loose ends. Vacuum the equipment as well as the operating table

How to troubleshoot the problem with a Ham Radio Station

It is possible to troubleshoot your Ham radio station in a variety of ways. These include information, control and power. If you are able to identify what is causing the issue it is possible to determine the root cause.

Problems with data arise due to computer interfaces that interface with radios or data processors. Internet controlled devices work with Wi-Fi or Ethernet. This means that you could encounter issues like network issues and baud rate issues, insufficient wire configurations, as well as protocol issues.

On the other hand pilot errors are the cause of radio control issues. equipment.

Solutions to Problems by using Power Supply

If you are looking to resolve issues with your power supply, be sure the connectors and cables are properly fitted. This is due to the fact that the majority of power problems result from connectors and cables that can be easily removed and then reconnected.

Review your control settings and make sure you have the correct connectors securely screwed. If you see a drop in voltage or a high SWR it is likely to be an unfastened connector. You may also look for short or open cables due to damaged cables or defective soldering.

Utilize a volt-ohm-meter to test the cables. Remove each end of your cable your device.

Problems to be solved with Radio Frequency

To avoid problems with radio frequencies, ensure that you have a license. After that, you can install an antenna, as well as any other devices correctly. Be sure to avoid congestion on radio waves, and also shield yourself from electromagnetic radiation.

Reduce interference and noise by keeping a certain distance from domestic equipment and employing filters. Reroute cables and make use of ground isolation to signal pathways. Also, repair electric fences.

How to troubleshoot the problems with your home and its surroundings

Here are a few methods to troubleshoot the problems in your house and the surrounding area:

a. Take a look at your cables and connectors.

b. Examine the controls for sound when you turn on the device but don't listen to any sounds. Check the audio-gain control.

C. Stop interference from other devices.

Controlling Interference with Other Gadgets

Interference with other devices could happen in various ways, for example, when the signal sent by a device is interpreted by a different system. This could result in an audible noise or broken connection.

Also, an overcrowded band and audio noise caused by speakers, the emissions of devices that are not monitored and radio waves that are overactive can cause disturbance.

To combat these interferences alter the position or location of your devices. Reducing the amount of people who are crowded on a particular frequency by increasing the range and the size that your antenna.

Controlling Interference to your Radio

It is possible to handle interferences with simple methods like

Rerouting cables and shifting equipment.

* Use arc snubbers and filters to the point of origin.

* Applying ground signal path isolators.

* Adding ferrite chokes , or shielding cables.

How to use the Kit and build your Station

You can make use of kits to construct your radio station. They include a variety of microphones. It is possible to start with the entry-level microphones, and then move to professional ones. Another option is software for broadcasting. A variety of broadcasting programs are available to broadcast live. There are also microphone processors that can amplify and equalizing to improve the audio quality.

How to build your Station from scratch

Are you aware that you could build the radio station of your dreams by scratch? Take these actions:

a. First step setting up the time and roles that your station has.

b. Include other program managers , or DJs.

C. Send an audio file or song.

d. Develop your initial radio show.

E. Make use of smartblocks and create playlists.

F. Add more content to your show.

g. Modify your radio page.

h. Start a live broadcasting.

Chapter 8: The Basic Technical Information Required For Ham Radio Station

Hobbyists or enthusiasts from all over the globe have discovered an avenue to achieve their goals in broadcasting, particularly in emergencies. The most fundamental technical skills you must possess include

A. Propagation is the process of allowing radio signals to propagate throughout the air before they hits antennas.

b. DX-ing - it means distancing.

C. Contests are intended to reach out to a number of people via the airwave.

D. Awards are given to hams who participated in contests.

It is. nets, or Network A meeting of radio operators across the airwaves to serve different reasons.

F. Rag chewing - basic conversations with amateur broadcasters over the radio.

Understanding the electrical Units and symbols

The electrical units and symbols they carry are:

Electrical Units S/N Symbols Significances

1. Resistor

Limits flux of energy

2. Capacitor

Double terminals that run into plates.

3. Antenna

A rod, wire or other device used for attracting radio and electromagnetic signals.

4. Earth or Ground

It is used as a zero-potential reference to measure the difference in current. determined.

5. Fuse

It prevents currents that exceed the limit of a certain amount from damaging wires and safeguarding electrical circuits.

6. Inductor

To store energy in the form of magnet or field.

7. Switch

This is used to disconnect the current while open.

Ohm's Law

Georg Ohm, a German scientist who conducted tests to prove the connection between current, voltage, and resistance. So, ohm's law says that "The flow of current in circuits will be directly related to its applied potential difference, and in turn inversely related to resistance within the circuit."

Analyzing the power of Ham Radio

Ham radio operations that involve transmitting at lower power and maximizing distance is known as QRP operation. However, in radio engineering, it is possible to analyse, convert, and control and mix diverse variations with AC energy or DC power with transverters. It is possible to alter the frequency range on which the transceiver is able to broadcast with a transverter as well as a the transceiver at the same time.

What are Attenuation, Gain and Loss?

Losses are equivalent to gains that are negative. So, losing 5dB is equivalent to an increase of 5dB. Additionally, 20dB of attenuation is equivalent to an increase of -20dB.

Furthermore both the benefits and loss of stages within a broadcast radio or a broadcasting unit can be combined into.

Using Bandwidth in amateur radio

The frequency range that is that are used to transmit information is known as the bandwidth of radio frequencies. The bandwidth refers to the frequency range that is that are received or transmitted in the event that the power is not zero.

Designs of Antenna in ham radio stations

Antennas with different designs are offered on the market. You can choose one based on the frequency of electromagnetic waves that travel between your transmitter and the receiver's device.

They include directional antennas that have a narrow beam to allow proper directionally-oriented propagation. Some examples include Yagi and parabolic that have different attributes and functions.

Another type of antenna is the Omni-directional antenna. It is able to propagate in any direction. Additionally, the third form is semi-directional antenna. It can be propagated in a specific shape. The definition is based on a particular angle.

What's Standing Wave Ratio (SWR)?

The measurement of the impedance that is generated by loads to the impedance of the waveguide or transmission line is known as SWR the standing-wave ratio. SWR Meter is an instrument that is used to measure stand-wave ratio.

Ham radio's SWR must be determined in the installation of Marine radio Business band, Marine radio, or CB. This ensures that the power that broadcasts generated by radio equipment is flowing properly throughout the antenna.

Use filters in Ham Radios

The radio frequency signal is utilized to control or permit certain specific frequencies, or radio signal. This allows you to filter out unwanted signals. This reduces interference from external devices.

How to track satellites in Ham Radio operations

Utilizing satellites or the "Birds," in the words of experts in broadcasting ham radio is a new technology that every operator must explore. However, you must track the satellites before you are able to utilize them.

Here are the best ways to track satellites to aid in your amateur radio activities:

1. Find out when it is within the reach from your radio shop.

2. You can monitor these satellites by using computers connected to the internet at the radio channels we have.

3. There are many satellite tracking software available included in shareware. You can also buy these programs for your private use.

4. A variety of amateur satellite companies have websites to track satellites.

The features from the Battery of the Battery Ham Radio

If your radio device is not connected to an electric grid, then you must make use of batteries to power your system. It is possible to use Go Box batteries with high range of voltages from 9 volts up to 15 Volts. They're mobile and can be used to power your equipment from wherever you happen to be.

Chapter 9: How To Make Contact

Ham radio stations can make connections with other hams via chats like Ragchews or on-air chats and contests.

Listening to contacts via your Ham Radio

Incorporating contacts or QSOs in your logbook is essential in order to communicate with them and listen to the broadcasts of their stations. If you're employing digital mode, make sure you be aware of what your computer is showing.

In thiscase, you tune in to the bands throughout the day to receive different frequencies, and then you watch or listen to conversations for the contact to chat with them.

Understanding how Radio Bands are organized

Radio bands are assigned from the International Telecommunications Union (ITU). These are the different ways in which bands are classified, for example medium frequency, low frequency, and high frequency.

Low Frequency

Low frequencies fall under that of Asian as well as European longwave transmission bands. They also fall beneath the broadcasting band for commercial radio. It is 2200 meters which areas ranging from 135.7 KHz to 137.8 kmhz.

Medium Frequency

This is lower than that band of maritime radio and the commercial AM broadcasting band which is 630 metres. It is located between 472 kHz and kHz.

High Frequency

High-frequency bands are

80 meters to 3.5 Mhz to 4.0 Mhz

60 meters - 5 MHz region

40 meters 7.0 Mhz to 7.3 7 MHz

30 meters 30 meters 10.1 Mhz to 10.15 MHz

20 meters 20 meters 14.0 MHz up to 14.35 MHz

17 meters 17 meters 188.068 Mhz to 18.168 MHz

15 meters from 21 MHz to 21.45 MHz

12 meters 12 meters 24.89 Mhz to 24.99 MHz

10 meters 28.7 MHz up to 29.7 MHz

How to Listen to the HF

It is possible to listen for repeaters that operate on a high Frequency by looking for and finding the repeater within your area via the directory or on the website. Indicate the output and input frequencies that the repeater uses. Use your radio to tune to its output frequencies of the Repeater. You can also apply listening to an input technique to obtain stations that transmit through the repeater. After that, you can tune your radio to FM signals.

How to Listen to UHF

To listen to contact calls on Ultra High Frequency (UHF) Here are the steps:

1. Find a repeater within your area or locality by visiting the website of the repeater or the directory.

2. Define the frequency of input and output that the repeater will use.

3. Your radio should be tuned at the frequencies of your repeater.

4. Utilize a method called listening to the input to detect stations that broadcast at the Repeater.

5. Your radio should be tuned exactly the same way you would tune FM signals.

How to Listen to VHF

If you'd like to listen to your contacts via Very High Frequency (VHF) Follow these steps:

Find repeaters near your area using the website of repeaters or the directory.

* You can find the frequency of input and output that the repeater uses.

* Set your radio to be tuned to the frequencies of the repeater.

* Listen to an input to the ham radio stations that broadcast on the repeater.

Tune your ham radio the same way you would for FM signals.

Techniques for Receiving Signals

On a typical ham-radio station, you will get signals in various ways, like

1. Through Antennas Through Antennas Antenna transforms electricity to radio wave.

2. Transmitters are an vital equipment in the radio station. They transmit radio signals over the air. The signal is also known as the Radio Frequency (RF).

3. Receivers are devices that receives signals from an antenna used by a transmitter.

4. Transceivers - In this instance the receiver and transmitter can be combined for the transceiver. This device also utilizes one antenna to operate.

How to receive SSB

If you're looking to receive a single Sideband (SSB) message Here are the methods to accomplish it:

a. Programm your device to be able to receive signals from the SSB signal.

b. After that, you have the option of selecting USB as well as LSB.

C. Select the largest SSB filter.

D. Continue to adjust this knob until get an SSB signal.

The voice will be a. Then, continue to tune until the voice sounds normal and natural.

How do I receive FM

FM is the acronym in for frequency modulation. It is employed in many areas of broadcasting ham radio because of its signal strength, its resilience and resistance to background noise. In this scenario frequencies of signal changes in accordance with its modulating data.

FM signals can be received on a variety of devices like handheld devices such as Walkie Talkies and other Very High-Frequency radios.

How do I Receive Morse Code

Here are some different methods to be able to receive Morse codes in amateur radio

* Configure the rig to be capable of taking Morse code by selecting that CW mode.

* Set the frequency to any of the frequency within the lower end of a band with high frequency like 20 kHz or 30 kHz.

• Set the rig up to utilize a wider filter when the rig is equipped with several filters.

Adjust or modify your tuning controls until you hear an Morse code signal that is broadcast on the airwaves.

How do I receive Digital or Data Modes

For receiving digital or data modes, connect your device to USB. You can also tune into any of PSK31's call frequencies. Look for different filters available. Choose one that is suitable for voice , such as the typical 2.4 KHz filters. Set up FLDIGI and configure it to receive PSK31. Switch off the waterfall display feature to access these modes.

How do I receive Digital Voice

In ham radio, various digital voices can be recorded in the frequency range of Ultra High Frequency (UHF) and Very High Frequency (VHF). One method to receive data or digital modes is by using the computer that is connected with the transceiver for High Frequency. Audio outputs from the transceiver or sound card are the audio input to broadcast.

Chapter 10: Understanding Qsos (Contacts)

Conversation of two stations on amateur radio is referred to as QSO. QSO. It involves contacts between amateur radio operators in the high-frequency bands of amateur radio and allocations.

There are many contacts that are available for Ham radio operation. If, as a novice, you know what these contacts are and how to create them, it is simple to find contacts once you start transmitting.

What's Chewing the Rag?

Chewing on the rag can be an extensive, formal discussion between radio hams waves. They might meet to debate on issues, have an event, or talk about various other issues. This could also be described as chewing on fat.

How to Connect with Other Operators On the Internet

You can connect with other operators on the internet by scrolling through ULS's ULS Advanced Search page. Select the drop-down menus that appear. You can also type with the boxes for text. They include Service Group, Amateur, State as well as Zip Code. Choose the state that is appropriate or enter Zip Code.

Status: Active, Date Type, Grant Date.

Hit to the Search button.

What is the difference between DXing and Contesting?

Another term used to describe competition is Radiosport. This is a sporting event that which amateur radio operators take part in. In the course of the competition, an amateur radio channel or operator attempts to reach as many people or other hams as they can.

DX-ing refers to receiving radio signals in remote locations. It also requires two-way radio communication with other radio stations located far away.

How to make your own Contacts

If you're interested in making contact on your own, you can watch how other stations conduct contacts to familiarize yourself with the procedure. Make sure to start with the first contact, as you'll be more comfortable in making further contacts via the airwaves.

Contact with Repeater

A Ham radio repeater contact can be described as an electronic gadget that picks up an amateur radio signal that is weak or low quality radio signal and re-transmits the signal at a higher level or at a higher level. In this scenario the signal can travel vast distances without losing its signal.

Contact HF

Amateur radio operators are able to make certain contacts or QSOs using high-frequency bands. The contacts made with the HF bands are known as rubber stamp contacts since they aren't able to hold long-distance conversations. However, they can help operators gain a lot of contacts within a brief time.

CW Digital Mode Contacts

CW refers to continuous Wave broadcasts. This is accomplished with the help of vacuum tube oscillators, which make a clear note. If the frequency is crystal clear then move on to ensure that you're not sending faster than you're capable of writing.

Then, you can program for the rate of the keyer. Make sure your mind is ready to go at any cost. Answer a 3x2 CQ. Listen and watch for responses for five seconds. If you don't get an answer, you can repeat the procedure.

What happens if you fail to make contact

If you don't reach out, you may continue to broadcast any information you believe is useful to the public until you hear back from someone.

How to Make an ongoing Contact

Breaking into an ongoing call in ham radio, particularly when you are unable to wait until the end of the contact to make a call to an Ham radio station. The interruption that you cause is known as breaking.

The way to address this issue is to wait until there's a gap in the conversation. After that, you can use Morse code and then send BK'. You can also say"Break. "Break" and then say your name and call number.

The Best Way to Run Your QSO

QSO is a simple term used to describe a conversations between two people who are amateurs. If you're looking to conduct a phone (or voice) QSO you have two possibilities. You can either call CQ or an operator on the radio who is calling CQ. CQ is general and does not refer to any particular person.

How to call CQ on High Frequency

To obtain a random number it is necessary to dial CQ. CQ is a generic phone call to obtain any contact. It is also possible to answer the person who is calling you. However, before calling CQ look for the frequency that is unoccupied and not being used by an amateur station. But, finding an unoccupied frequency could be difficult if you have an overcrowded band.

Chapter 11: Involving In Casual Operation

Ham radio is a medium where you can participate in informal activities like engaging in casual contact, conversations or other interactions with old and new friends via the airwaves.

How to Operate FM with Repeaters and Simplex

To use FM using a repeater as well as simplex, you will require an transceiver capable of transmitting using the frequency of the repeater's input. Each band has the frequencies of the output and input will be divided according to a set amount. The process of dividing frequencies is known as offset.

The simplex repeater may also be known as a parrot repeater. It is made to work with both a transceiver and voice recorder of very short duration. Thus any magnets that the receiver has for a specific amount of time are recorded. The recording typically lasts thirty minutes or so. The message recorded is played back to the transmitter at the same frequency.

What exactly is Repeater Basics?

The basic repeater functions in amateur radio consist of an FM receiver on one frequency, and an FM transmitter operating on another frequency. It usually happens when there is interconnection between bands. If one receiver is able to detect signals and the transmitter is then activated and then broadcasts the needed information.

How do I use Access Control Tone

Access control tones comprise DCS as well as CTCSS. CTCSS is only analog, but should you decide to build an entirely new radio the process will be much easier to manage. But, you will require an electronic processor to process the DCS signal. Additionally, DCS can be activated accidentally.

The CTCSS tone have similar sounding with the power grid in your area. CTCSS is a combination of decoding and programming while Tone is a coding.

Using Miscellaneous Repeater Features

This feature requires the use of a transceiver which can broadcast on the frequency of the repeater's input. Other repeater features can be able to receive signals on the output frequency. A predetermined amount of time is utilized to divide the output and input frequencies of each band. The process of segregating the frequencies is known as 'Offset.'

How to Set-Up Your Radio Station

Here are some ways to establish for your station on radio:

a. Programming Your station's timezone, station and the roles of your users.

b. Include DJs or Program managers.

C. Now, you are able to upload audio.

d. Create you're your first radio show.

E. Create playlists and then apply smartblocks.

F. Create content for your show.

g. Design a custom radio page.

H. Engage in broadcasting live.

How to Utilize Simplex

Simplex is a method by which radio stations that are on the same frequency communicate with each other in a direct manner. The stations transmit and receive at the same frequency, one after the other. In this scenario there is no repeater that is in between them.

Managing Digital Voice Systems

Digital voice modes form part of digital technology which gives better performance than analog ones. When signals are at lower levels they can be copied and they provide better performance and efficiency in signal transmission.

Digital voice systems are Fusion and System Fusion, DMR, and D-Star.

High-Frequency Digital Voice

Different digital voice modes can be utilized in High Frequency. If you have a computer one can directly connect to the transmitter on High Frequency. In this scenario, audio will be broadcast from the sound card as well as audio output. This will then drive an audio signal and transmit the information as it is.

Digital high-frequency voices comprise D-Star as well as FreeDV.

Ultra High-Frequency Digital Voice

Ham radio is a medium to use Digital voice channels. The frequency used for making use of the voice modes is VHF. You must select the one you prefer from a list from digital voices. There isn't any norm for FM, AM SSB as well as FM bands.

These digital voice technologies used for Ham radio operation include System Fuse, D-Star, and DMR.

Ultra High Frequency Digital Voice

The digital voices of Very High-Frequency bands work similar to Ultra high-frequency band mode. However, you must select the voice mode you prefer which can be synchronized with AM bands, FM bands, bands or Single Sidebands.

Utilizing Digital Repeater Networks

Digital repeater networks consist of an FM receiver that operates on the frequency of an FM transmitter that operates on another frequency. The transmission usually occurs in the same radio frequency linked in such a way that the receiver receives an indication when the transmitter is controlled. It then broadcasts the signal recorded or registered within the system.

The best way to chew the Rag

A long, pleasant conversation is the meaning of the phrase "chew the rag. You can chew your rag by calling your contacts in your area or even distant stations to have conversations. The station you call will start the conversation with you. It could be a live roundtable discussion of specific events.

How to Chew the Rag

If you'd like to speak to any station, you may dial CQ. By calling CQ, you say that you'd like to talk with the station from a distance. Based on the social aspects of the time for contact most people prefer weekdays for the ragging.

However, the majority of amateur radio stations participate in ragchewing during weekends, though a variety of social activities are conducted during the weekend. You are able to choose a more suitable time for you and your group.

How to Chew the Rag

Rag chewing is evident in different frequencies across bands. The bands all have similar structure. For frequencies below 30 MHz, one will see Morse codes as well as digital modes within the bottom third. Additionally, voice modes can be located in the upper two-thirds.

You can view different kinds of contacts in a specific frequency band.

Recognizing the Ragchewer

You will be able to recognize an Ragchewer when the people who have been in contact spent a good amount of time talking with radio waves. Then, break into and join in the group of rag chewers. Check you're radio station has a powerful signal. Take what they're doing and implement it in your station.

Chapter 12: Participating Public Service Operations Public Service Operations

Ham radio operators may participate in public service activities through providing data and voice communications during disasters. They are able to operate in medical centers, police stations fire stations, as well as auxiliary command posts. They are also found at evacuation sites as well as emergency shelters.

How do you join an Public Service Organization

Public service organizations are comprised of local volunteers in the region. They are governed by the city or county. Each participant must have a valid license as an operator of ham radio. They must also be interested and willing in giving selfless service to humanity when emergency situations arise.

How to Find the Public Service Group
Public Service Group

It is possible to find an organization that is public service within your local area by searching for one that normally hosts Amateur Radio formal messages. There are numerous such organizations or networks that are available both locally and nationally. They are mostly affiliated to the National Traffic system in different regions and areas like that of ARES, SKYWARN, and RACES.

How can you offer your services for free?

If you are a ham radio operator You are able to volunteer your time and talents through joining ARES. These are volunteer groups that are organized in a city or in the country for emergencies.

To join you must possess an active ham radio license and be able to share your expertise and knowledge to assist the community during emergencies. The person who organizes these groups serves as the disaster coordinator.

How to prepare for disasters and Emergencies

Ham radios are equipped to be prepared for natural emergencies and catastrophes such as floods, earthquakes and war. They can also be used to prepare for terrorist attacks. These are because emergencies can disrupt communications channels.

They must also be formally certified by the regulators such as the FCC before they are able to begin operations. They are part of Amateur Radio Emergency Services (ARES).

How to operate in disasters and Emergencies

In the event of disasters and emergencies Ham radios are part of groups who manage communications. Amateur radios also participate in local emergency groups. They also are part of regional traffic nets.

Similar to that the ham radio operators participate in state emergency management activities in the level of state. They offer voice and data communications, and are also helpful at remote post of command.

Ham radios are a valuable tool for fire and police stations as well as medical facilities, shelters in emergency situations, on public work areas and evacuation centers.

How to report an incident

If you'd like for an official report of the incident using an amateur radio user, just dial 911. If the operator is responsive to your call, you must mention your name and declare that you are informing them of an issue. Follow whatever the 911 operator instructs you to do, such as "Stay at the end of the line." Additionally, make sure that other repeater users remain in the call.

Tips to make Distress calls

Distress calls must be made by calling MAYDAY-MAYDAY and indicating your number. It can be done locally by using either UHF or VHF. You can also accomplish this by using distress calls to high frequency.

How to respond to distress Phone calls

To reply to distress calls, you can use the Distress signal. It's MAYDAY. Repeat it three times, and then repeat the word "THIS IS" at least once. Then, repeat the name for the ship or the area in need of assistance three times. Make sure to mention your call sign as well as the number for registration of your vessel at least once. After that, you can say "MAYDAY" as well as the title of your vessel, or location once.

How to monitor weather and forecasts?

Amateur radio communication is one way the Office of the National Weather Service Forecast interacts with emergency management agencies and spotter groups.

They've been providing support by way of communication to SKYWARN Storm spotter programs using special communications.

How to participate in nets

Participate in nets by logging in to Ham radio networks. Contact the NCS and make sure you register your call sign, your location and the status. It is important to read the guidelines of the NCS well before signing up to the Nets.

How to Check In and out

You can log into a web site by registering your phone number along with your location and the status of your call with NCS. The most important thing to strive for is to listen to the instructions of NCS clearly. Try to understand the rules. However, if you're not a regular participant on the web, you need to be patient until NCS invites guests.

Ways of Exchanging Information

Ham radio operators employ a variety of methods of communication to communicate. The Single Sideband (SSB) and Frequency Modulation (FM) are among the more popular modes of voice broadcasts. It offers high-quality audio signals , however SSB is preferred for long-distance transmission.

What is tactical call Signs?

They are used to aid in emergencies and tactical networks. It allows you to identify the mission, location and other functions of stations.

Utilizing Digital Message Networks

Digital message networks operate by transmitting modes via the microphone input on the radio, and then adding audio to it. This is achievable using the analog circuits such as FM, AM, D-Star and SSB. These are high-speed, data-only modes.

Sending email via Radio Sending Email via Radio Winlink

Amateur radio operators are able to communicate and get Winlink emails with the WinMOR mode, either digital or PACTOR on their high-frequency radio. It can be done via the packet or packet mode on VHF frequency or UHF frequency.

However, on high Frequencies you can transmit Winlink via WINMOR by using a software or a sound card. There could also be separate communications processor that is compatible with the range of PACTOR digital modes.

What is What is AREDN?

Amateur Radio Emergency Data Network is the entire definition of AREDN. This function provides an opportunity for radio operators to create a an ad-hoc data network with high speed that is useful in emergency situations and for broadcasts based on service.

Chapter 13: Understanding Digital Definitions

A few examples of digital terms you'll see on amateur radio include

A. Packet radio transmits information in packets to receiver stations.

b. AMTOR - means Amateur Telex Over Radio (AMTOR).

C. WSPR - software developed by Joe Taylor for transmitting weak signals between stations.

D. The PACTOR is used to perform frequency Shift Keying (FSK) modulation using the HF.

E. RTTY - Radio Teletype - one of the first digital formats that were used to create large teletypes or the teleprinters.

For more information, click here. PSK31: This is Phase Shift Key (PSK) transmitting data at a rate at 31.25 bits/second.

G. the WSJT mode - which involves weak signal communication with K1JT. Digital communication is a method employed for weak signal communications.

What exactly is PSK31?

This is a reference to Phase Shift Keying (PSK). It's also known as BPSK31 or QPSK31. These are familiar computer-audio-card-generated radioteletype mode. It is employed by radio operators in ham radio for managing keyboard-to-keyboard chats on the HF band.

The MSFK mode is available on Ham Radio

The MSFK mode on ham radio include Morse code that is the first digital mode. It's a radio telegraphy that is designed for machine-to machine transmissions. Analog voice which is an analogue modulating radio transmission that includes the modulation of phase, amplitude and frequency through analog waveforms.

FSK, Radioteletype (RTTY) along with FSK

This is done by using tones for transmitting digital messages to ham radio stations operating in high-frequency bands.

Fast and Slow Modes for WSJT

This program is computer-based to allow weak signal radio communications between radio operators of ham radio. Joe Taylor was the initiator of this program. It is now available as open source.

DX-ing - Distant Stations

The process of being able to detect and receive signals from radio stations that are far away is known as DX-ing. It is also referred to as Distancing.

DX-ing in the Shortwave High-Frequency Band

Any radio band that is in the frequency range between 1 MHz to 30 MHz is called low frequency shortwave. It includes Ham radio and CB and listening to shortwave.

DXing in UHF Bands

UHF bands are intended to be used for line of sight communications, also known as short range transmissions. In this scenario it is possible to consider distancing broadcasts from stations located 50 from 100 or 50 miles.

Bands of microwave and ultra high-frequency bands are also designed to provide global transmissions from earth to moon across radio broadcasting stations.

DX-ing in VHF Bands

DXing in VHF bands is performed at frequencies that exceed 30 MHz. The difference from the high-frequency bands is the fact that radio signals do not be reflected off of the atmosphere. In this instance there is a different method for transmitting or receiving radio signal.

You can look at the lower part of the radio band with an antenna beam small enough to find the appropriate frequencies.

Automatic Link Establishment (ALE)

Automated link creation is the main principle behind the creation and maintenance of radio communications. This is possible with high-frequency radio. Amateur radio operators form a community. operators has been operating on the airwaves for seven years.

Tracking and APRS

APRS can't be used to track vehicles, like vehicle trackers , but this is the case with the ham radios. If there is something that is occurring right now that is of concern to you, it will be visible on your radio's mobile device to draw your attention and prompt a response.

Spread Spectrum and Broadband Hamnet

It isn't an option to hook up your Ham radio directly to the internet. However, you can add internet connectivity to your radio by using a USB connection and computer by using a dedicated interface.

This allows you to control the radio with the remote control.

Chapter 14: How To Participate In Radio Contests

Amateur radio contest can also be called radio sports. Ham radio operators take part in this contest. The competition involves communicating with other radio stations through using an amateur radio network. This happens within a certain time frame.

How to Run a Contest

Here are some ideas for running the radio contest:

Each radio band is equipped with the type of communication which can be used.

Every radio band is equipped with the type of information that must be exchanged.

* The nature and quantity of contacts you had during the contest determines your score.

Your score earn is used to rank stations.

The results are then published in magazines and websites from the sponsor.

Get advice from the winners of an Event

Some of the tips include:

1. Keep short communication.

2. Contact them at a time which allows you to listen to the caller and then respond.

3. Pause for a few seconds between CQs prior to making another call.

4. Continue to make contacts even if already have contacts.

Pursuing Awards

If you've chosen an award for contest You must keep the patience and perseverance.

What is the best way to apply for awards?

If you'd like to be considered for an award on Ham radio, you are able to apply for it regardless of age, experience, and the level of your license. Be aware of the various categories of awards and determine which one that you could apply for. There are RSGB HF awards as well as RSGB UHF and HF awards.

RSGB HF awards - helps you move in one step to the next.

RSGB UHF and HF awards allow you to utilize frequencies that are dependent on certain methods of propagation.

How to find Special Awards and Events

One of the most unique ceremonies and awards that have been held in amateur radio's history is the DXCC highly coveted award.

How do I log or record Contacts

It is essential to record contacts in order to record your interactions in the event there's an indication of interference from other stations. It is also important to keep a record of prior contacts to be able to fill in your DX QSL card. Also, it can help you keep track of all the people and locations you've interacted with.

Chapter 15: Understanding The Morse Code And Using Morse Code

Morse code can be described as a technique employed in telecommunications to encode text characters. It is a standard series with two different durations of signals. They are also referred to dots and dashes. Another name used to describe Morse code is Morse code is dahs and dits. Samuel Morse is the originator and inventor of the telegraph.

Move to the left whenever you hear a dit or turn to the left when you hear the sound of a dah. You may hear a dah dit dit. It means dots, dash, and dot. Write the letter D on your copy paper. Then, move back to wait for the next letter.

Start by introducing Farnsworth

When you're looking to transmit code, you can make use of Farnsworth. The reason for this is that during a conversation, long words can lead to a an inability to concentrate.

Copy the Code

To write Morse code specifically in your memory, just shut your eyes, and repeat the words in your head as you learn them. The memory becomes quicker until a certain point. If you're slow and you are not able to hear the letters, they are also going to slow down in your mind.

How to Create Code Contacts

Examine the ways other hams interact and study their methods. Next, start your first contact. Amateur radio operators typically change the mode of their radios from transmitting signal to receive signals.

Utilizing the QRP (Low power) and Portable

QRP involves transmitting at lower power while increasing your range of operation. QRP is a unique attempt at radio transmission that gained acclaim during the Nineteen Twenty-Twenties (1920).

The Intricacies of QRP

The transmitter of QRP often faces obstacles when trying to maximize their areas of coverage. They generally limit the radio frequency output capacity to 5 or less watts when it comes to CW operation. It also allows 10-watts of power or less in SSB operation.

How do I Operate a Portable QRP?

Operators of QRP might adopt different strategies like improved antenna systems and antennas improved capabilities for operation, and a variety of special settings to enhance their capabilities. This can help them keep good radio contact.

ARDF Find Direction

ARDF simply means amateur radio direction finding. Other names for ARDF include radiosport, radio orienteering or radio fox hunt. It is a sport of racing in ham radio that combines radio direction finding, maps and compass abilities of the orienteering.

Activities of ARDF utilize RF in 2-meter or bands of 80 meters.

Chapter 16: Understanding Some Terms In Ham Radio

In the field of ham radio there are a variety of terms used to describe a wide range of definitions. A few of these are

A. QSL signifies, "I confirm reception." This is used to confirm contact.

b. 1973 - Used in Ham radio communications to express best wishes. It is recommended to be used at the end of the contact.

C. CW - CW transmission using vacuum tube oscillators that produce pure notes. Ham radio operators utilize solid-state and microprocessors to allow different communication modes like digital data formats, voice, CW, and image.

What are Q-Signals that are spoken?

Spoken Q-signals are a set of radio communications that are shorthand employed by radio operators who do not have that same language. They are used to aid in interactions, thus making speech speedier. These signals were developed using old telegraphy notes. It can help to reduce time and enhance communication between contacts using radio waves.

Different Antennas

There are three kinds of antennas: Omnidirectional antennas, semi-directional antennas and the directional antennas. Antennas with the ability to travel in all directions are known as Omnidirectional antennas. Antennas with the ability to travel in an axis of field are called semi-directional antennas, while the ones that can be directional are defined by a specific angle.

Utilizing Feed lines

The conduit for radio frequency between the antenna and radio is known as the feed lines. Also called the transmission lines. The energy you generate is sent to the antenna via the feedline. In the same way, every signal gathered by the antenna are transmitted to the radio via the feedline.

What is Antenna Tuners?

Antenna tuners work in many ways, such as in the feeder that is close to the transmitter, or at locations where antenna elements connect to the feeder. This is because it ensures that the impedance system of the antenna will work with transmitter or receiver. So the best performance can be realized.

If your SWR exceeds 1.5 it is possible that you require the aid of an antenna tuner. This can change based on the frequency that you're operating at. The device is helpful for receiving signals. If your radio is receiving 50 ohms in the feed the signal will be higher quality. You can use the noise floor or received signal to tune your antenna system.

Analyzing Grid Squares

Grid squares are the simplest way to use shorthand for analyzing your location anywhere on the earth's surface. It is simple to communicate it on radio waves.

Two numbers known as "the square" serve to define this grid-like square. Additionally, two letters referred to as "the field" are utilized to describe the square.

The grid's locator square is known as the Maidenhead Locator System. It is a method of geographic co-ordinate that relies on a six-digit number. It is commonly utilized by radio operators to establish a rough position on the planet.

Solar as well as Geomagnetic Activity

Solar activity is visible in the form of xrays, such as those that flare up in September. Radio signals of high frequency can be absorbed by the Ionosphere as a result of increasing solar energy that is that is released. The solar energy can be valuable for amateur radio users.

Geomagnetic activity, however, is important for propagation of very high frequency. This magnetic field is located in the direction of the poles. It's created to create aurora that can reflect the signals of Very High Frequency. This way it will increase the likelihood of propagating transmissions with long ranges.

Connector Parts

Ham radio stations you will find a variety of connector components that are essential to operate the shack. They include shielded phono connectors microphones, the common radio frequency (RF) connectors. Other connectors include BNC and N as well as DIN as well as UHF.

Chapter 17: Tips For Enthusiasts As Well As Hobbyists

This chapter provides suggestions for amateurs about how to operate their radio shacks, and how to succeed.

Know How It Functions

Radio stations that are amateur can utilize different frequency bands across the spectrum. They are governed by the Federal Communications Commission allocates these frequencies to amateur radio operators. Ham radio enthusiasts can operate above the AM band, and into the microwave area within the gigahertz band.

Keep the Rules in Mind

Follow the rules for radio broadcasting to ensure you are broadcasting as long as possible. Certain of these rules cover

1. You will need to obtain the necessary license to operate radio stations.

2. It is necessary to pay a token cost and take the tests before the ticket is handed to you.

3. Always end your call at the conclusion of conversations by stating your number as well as the stations you're calling.

Pay attention to details

Learn what you must do in any situation. If you don't follow the rules, it could lead to serious issues which could result in the loss of your license as well as the ability to operate the Ham radio station.

Continuously improve

Try to improve your service by using radio waves. Get more sophisticated equipment to improve your offerings.

Find out What You're Not Sure About

There is no such thing as an isle. Instilling the habit of continual learning will allow you gain new ideas about how to spread and run your station. Inability to study and gain more strategies and ideas can cause you to be off the grid. Keep positive relationships with fellow hams from your area.

Always be ready to help others and receive assistance too.

Inspire young and future radio enthusiasts and hobbyists within your community to participate in broadcasting amateur radio. It is possible to do this through joining clubs in which members share ideas and give advice about the best actions you should take in any moment in time.

The practice will allow you to get better at it.

Radio broadcasting by amateurs should be a constant and constant practice. As you get more involved in the activity, the more you'll learn more techniques to enhance your radio signals and efficiency.

Chapter 18: Ham Radio Basics

How do you define "ham radio?"" This is the term used to refers to amateur radio. Ham radio users make use of radio frequencies to exchange various non-commercial messages with fellow hams for various reasons like discussing the aftermath of an emergency (like flooding, hurricane and so on) or simply talking about common interests. To become a ham radio operator, you must take a test (which we'll discuss in greater detail later) to prove that you understand how to utilize "Amateur Bands," which are the radio frequency that hams are permitted to make use of. There more than 2 1/2 million hams in the world.

Ham radio first came into existence in the latter half of the 19th century, however the concept of ham radio as we understand was something that happened in the 20th century. It was the FCC

(Federal Communications Commission) realized that the ham radio was a viable option in times of emergency, and came up with an Amateur Radio Service. The main reason for this was the Titanic catastrophe which is the reason it was approved in 1912 by Congress through"the "Radio Act" of 1912." In contrast to regular radio, radio hams offer the opportunity for two-way communication and group conversations. If the Titanic was equipped with a radio ham, they could have received assistance much quicker and many lives could be saved.

There are three major kinds of ham radio interactions:

* Ragchews

This is slang used to describe just talking which is also known as "chewing at the rod." There's no

There is a specific purpose behind ragchews. Hams can talk about anything which is why it's an

A fantastic way to meet people across the globe.

* Nets

This is a contraction for networks. These are interactions that are scheduled using a

Specific purpose. They are also emergency service nets, that are used for emergencies.

Other communication methods (like telephone service) cease to function;

technical support that are usually designed to address questions from

new announcements from hams and bulletin boards. This is the term used when a computer system

It monitors a frequency when people connect, send and receive messages. Many

Hams often use bulletin board nets even when they do not do not have Internet.

* Contests

This is a great game. Hams will be competing against one another to see who is the best.

Reach the furthest station to have the highest number of contacts. There are other ways to

Events where hams get an time frame and are able to determine who is the best

Make the most contacts.

Why should I study ham radio?

Aren't convinced that you should know how to utilize the Ham radio? There are a variety of scenarios in which the ham radio could come into play:

* Traffic

Many times, you don't even know there's traffic until you're inside it. With a hammer.

radio listeners can hear the movements of other drivers who have Hams.

Real-time and avoid the real-time, and avoid the.

* Weather emergencies

In severe storms, bad weather can cause the loss of electricity, leaving no idea of what's happening.

What's happening. Ham radios that have batteries are the only way to be sure

to communicate. After the storm Sandy the towns were left without power shut off

during the past few weeks. a lot of people required emergency assistance. Ham radio

Operators are the sole ones to talk. 700 of them were able to communicate.

around 200 shelters that provide services to the homeless.

* Terrorist attacks

The 11th of September, 2001 ARES, a ham-radio organization in the United States, was formed. ARES was formed.

communication five minutes after the plane was sunk into the north

tower. They worked for hours and provided constant

information. This is only one instance of how civilian operators are able to

Assistance during a time of need during a.

* Power grid interruption

The Survivalists (or "preppers") often speak about the possibility of a power grid breakdown. This is

When a major event occurs such as an natural disaster or a the military's attack

The power grid in general is caused can cause the entire power grid to fail. Computers and cell phones

It won't work, and the only alternative is radio. Utilizing ham radio

individuals will be able arrange, trade products, and everything else.

essential to ensure that society is running smoothly.

* Education

There are a variety of ways the ham radio can assist in education. Making a ham radio stations or fixing an existing radio offers the opportunity for hands-on learning for students whether they're at high school or in an adult course. Communicating with someone from around the globe can be a wonderful method to master a different language, while also connecting and communicating about space stations. International Space Station is especially engaging for math, science engineering and science students.

Frequencies

Before we get to the ins and outs of Ham radio, it is necessary to understand Radio waves as well as frequencies. The first thing to understand is that radio waves are electrical wave that oscillates (like the string that vibrates in the guitar) which is reproduced through an antenna. Radio waves have distinct frequency. The radio receiver which the same as a radio in the way we think of it, has the ability to be tuned to a particular frequency, for instance, 97.1 KTCZ-FM , which is in Minnesota. The letters are given to the FCC. The 97.1 refers to the FM frequency that measures in megahertz. Megahertz is the equivalent of million of cycle per second.

For higher frequency wavelengths, these cycles are more efficient. This is why they have a shorter wavelength. Low-frequency waves, slower time to finish a cycle are longer in wavelength.

All FM stations operate between 88 and 108 megahertz. This is known as "band.

"band." Am stations in all markets are restricted to a range of 535-1700 Kilohertz. What is the reason why AM and FM are different in measurement? The reason for this is that AM radio was first introduced and all the frequencies that were assigned to different stations/broadcasts were very low. FM radio was introduced in 1939, however it was not popular until the 1960's, which means that the frequencies of FM radio were higher.

If you're considering the ham radio industry, you're granted certain bands to operate in. Ham radios can operate above AM broadcasts into the gigahertz band known as"the "microwave area." The majority of ham radio operators operate between 1.6 to 27 millimeters. Since lightwaves generated by the sun as well as radio signals are electromagnetic, the time of day can affect specific frequencies. If you are looking to connect with someone distant during the time of day, 15-27MHz is a great band and when night arrives the frequency changes to 1.6-15MHz. The bands you're allowed to use depend on the type of license you have. Below are the band bands and the purpose for which they are used:

* Medium frequency (MF) which is located between 300 kHz to 30 MHz. It includes AM broadcasting as well as a band for Hams.

* High Frequency/Shortwave (HF) ranges from 3-30 MHz. This includes shortwave broadcasting, ship-to-ship and ship-to-shore, as well as the military and nine bands used by Hams.

* Very high-frequency (VHF) that is between 30-300 MHz. This includes radio broadcasting on FM, security, commercial mobile radio as well as aviation and military and three bands for Hams.

* Ultra high-frequency (UHF) is between 300 MHz and 1 GHz. This includes cellphones, military/aviation, public

security and two bands to accommodate Hams.

* Microwave - any frequency higher than 1 GHz. It includes GPS as well as Wi-Fi wireless networking as well as digital wireless phones. military and eight bands to Hams.

In total you can choose from 26 band for amateurs. Bands are named using meters. For instance the 3.5-4.0 Band is known as the 80 meters band. The frequencies can differ slightly based on region. If you plan to use your radio for communicate over distances of a long distance, you need to make use of HF bands at a minimum.

Are there different types of Ham radios?

Once you've mastered the frequency range and how ham radios operate we can discuss the various types of Ham radios. There are four kinds of ham radios:

* Base/fixed-station radios

They are more powerful and feature the highest power. They are intended to remain

all in one place, and is able to be altered with a wide array of different accessories. They are generally more expensive and difficult to operate dependent on the complexity of the configuration is.

* Portable radios

Portable radios are the smaller versions of fixed radios. They are designed to

can be quickly set up in temporary locations that aren't equipped with either power or space.

But they're not compact enough to carry around on an everyday basis.

* Mobile radios

Truckers use mobile radios a lot. They're fairly easy to set up and operate.

It is useful for those who are frequently in the car. Mobiles and portables are often

Used to refer to the same thing.

* Handheld radios

These tiny radios easily be carried around in backpacks and are the basis of the new

Hams make use of. They're the cheapest however, they're also the least robust

When compared with the other types.

So, which one should you choose? In the next chapter, we will dive into the four kinds in greater detail and help you determine which type you should purchase.

Chapter 19: Selecting A Ham Radio

The choice of a ham radio may be a challenge if you're just beginning to it. If you decide to choose a permanent, fixed rig or something that is more portable such as a handheld is dependent on the things you'd like to achieve with your Ham radio, the amount you can afford and the level of your expertise. These are the pros and cons of all radio type.

Fixed Hams

These radios have the best features, and also a lot of power. Fixed ham radios that are high-frequency are commonly referred to as "rigs," have better receivers and are generally more user-friendly as compared to other radios. They were once the most expensive type but nowadays, you can buy new and used rigs at reasonably low cost. There are three types of rigs that you can pick from:

* Analog

They are the hams that have been around for a long time. They employ mechanical equipment and employ less energy.

greater power than digital, and needs computers. Analog hams also have

It has been proven to be more stable. This is vital during emergency situations.

the power could be cut off. But, if the weather is bad,

Analog systems are more prone to crackles and fuzz. Because they're

The old-fashioned way is the only way to go. There's no revolutionary technology being developed.

With them, you can expect features such as push-to-talk as well as encryption

capacities, that's all there is to the extent of it.

* Digital

It's the most modern Ham's system. It's simpler to use, particularly since

The majority of people are accustomed to computers nowadays. They consume less bandwidth, and

are more efficient. Additionally, they're more secure in conditions.

Changes, but don't make those nebulous changes. In the end, they are less costly and are able to make

plus, it's not a huge amount of accessories, compared to more, and with fewer accessories than.

* Digital/analog

These rigs utilize both analog and digital signals. The analog signal is the most basic.

acts as a backup any time the digital system goes down is to ensure that you have the very best both worlds. There's even more possibilities in projects that involve Wi-Fi and

other wireless advancements. Due to this, they're often the most popular.

expensive.

If you're prone to staying in your home for extended durations, live close to a radio tower that has the capacity to pick up an antenna and you have the space to install your radio and a fixed ham could be a great investment. If you're looking to make certain that your radio is equipped with sufficient power to connect with other hams who are located far away, that's another reason to consider getting fixed.

Mobile/portable hams

Mobile radios are less powerful in output power than fixed Hams. They range from 34 to 50 wattsand offer decent reception and transmission. To power them they require an external DC as well as an externally-attached antenna. Regarding benefits, they all are designed for those who travel often. Ham radios have been in use in their vehicles from the early 1920's. If you are involved in a crash and don't have cell reception and an ham radio on your mobile could be the lifeline.

Mobile/portables are built to be able to fit in almost all vehicles which includes trucks. you simply put the antenna in your vehicle. It's an excellent option when you do not have enough space to install an antenna in your home. The one thing that's consistent in mobile radios is their buttons aren't big, and you should be comfortable with the radio when driving. If you're not in a position to be, it's risky to operate it while driving.

Handheld hams

Handheld and mobile radios are alike, however handheld ham radios are the most compact and most affordable choice for those who are just beginning. On the plus side they're extremely useful for travel and, as technology advances and they're equipped with many features such as GPS tracking as well as regular AM/FM. Also, make sure that your radio is equipped with an internal memory that allows you to save the top frequencies and also an easy push-to-talk feature.

On the other hand the size of their devices means that they aren't as powerful and therefore do not have the same reach. If you're new to Ham radio but don't want to invest too much in the beginning, a handheld is an excellent choice, and you can carry it wherever you go.

What is the power?

In terms of power, let's explore the power of ham radios. The power of a ham radio affects the distance you can talk. Small handheld hams may only have a few milliwatts while the larger ones you can find for desktops can reach 1500 watts.

Since fixed radios can have such the highest watts they are the ones with the highest potential, and are able to access distant frequencies.

Mobile radios may have up to 10-15 watts. They usually require any kind of power source, such as a car or any other DC device.

Handheld radios require batteries that provide between 3 and 5 Watts of power. They can be powered by triple-A or double-A batteries. To ensure that your battery is in top shape keep your batteries in a cool place. As batteries lose energy, the ham radio will become less efficient So, always carry new batteries whenever you travel using your radio. If you opt for an handheld radio, examine whether the radio is equipped with rechargeable batteries as well as charger, if it's important to you.

In the majority of radios, particularly the latest models, you can have the ability to control the power settings. So you can select a low-energy option to reduce power consumption.

Do I need a dual, mono or multiband radio?

The name implies that the mono-band radio operates on one band. This typically falls in the 144-148MHz band, or the band of 440 MHz. Dual bands work on two bands, while multiple bands work with three or more. What exactly is that?

Let's suppose you have handheld radio with dual bands. This means that you are able to monitor two frequencies simultaneously. You're at a Ham radio gathering and all attendees are using their own radio. There is however one frequency that only staff members can access and you're an employee. You're able to observe both the regular frequency employees who aren't staff members use as well as the frequency used by staff members simultaneously, as you're equipped with an a dual-band transceiver. Then, you can select which frequency you'd like to use when talking back to the person you're talking to.

It's also helpful to have a variety of band choices when bands are full. In an emergency situation, particularly when a band is crowded, it can be a hassle and can be confused. It's easy to see why the dual band option, at the very least, is the best option, but this isn't always the situation. Multiband and dual band radios are more expensive particularly when you're starting out, it might be better to get a feel for ham radio with a mono-band radio that isn't expensive. If price is your primary issue you can locate inexpensive dual-band radios because there are more manufacturers that make them than mono-bands.

Where can you get your radio

There are numerous sites and shops that sell Ham radio, and it's essential to find out which brands you should consider. Icom, Alinco, Kenwood and Yaesu all produce excellent radios for hams. In regards to making a choice between fixed rigs, handhelds and more it's beneficial to check out reviews from other operators of ham radios.

If you're looking to buy a used radio, the Ham Radio Outlet (http://www.hamradio.com/used.cfm) has a used equipment section, as does eBay, though buying used equipment over the Internet always has its risks. For a list of hamfests (a kind of market that consists of ham radio operators), Google "hamfests in my local area." It's not uncommon to locate people who are who sell used equipment.

If you are buying your first radio simple and easy operation is the way to start. You don't want to invest a an excessive amount of money for one you're not sure of how to operate and may not be as good as you thought it would be, as the vendor claimed. Consider what you'd like for your new radio and then find the right radio to meet those needs.

Conclusion

Amateur radio stations are licensed by the FCC to handle emergency transmissions during disasters such as flood and other natural hazards. You can set up your radio shack anywhere that is convenient for you such as indoors on a desk at the corner of your apartment, outdoors under a shade for temporary transmissions during contests or events, and mobile transmissions using vehicles or hand held radios like walkie-talkies, etc.

You need to connect a suitable antenna to boost the reception of signals on your station. Various antennas are needed such as directional antennas, semi-directional antennas, and omnidirectional antennas. Choose anyone you need based on your transmission needs and the power of your transceivers.

You can build a remotely controlled station, if you want to operate outside your home. In this case, you can run the station from its main location but another person will be allowed to handle the operations for you. The station can function using a remote system or an automatic control.

You can also use a link to operate your station away from home. You can use devices such as laptop or smartphone to control your radio station while you are in the hotel, at the beach, or other recreational places.

www.ingramcontent.com/pod-product-compliance
Lightning Source LLC
Chambersburg PA
CBHW071838080526
44589CB00012B/1032